T0176697

ESSENTIAL PRACTICES FOR CREATING, STRENGTHENING, AND SUSTAINING PROCESS SAFETY CULTURE

PUBLICATIONS AVAILABLE FROM THE
CENTER FOR CHEMICAL PROCESS SAFETY
of the
AMERICAN INSTITUTE OF CHEMICAL
ENGINEERS

ESSENTIAL PRACTICES FOR CREATING, STRENGTHENING, AND SUSTAINING PROCESS SAFETY CULTURE

CENTER FOR CHEMICAL PROCESS SAFETY
of the
AMERICAN INSTITUTE OF CHEMICAL ENGINEERS
New York, NY

Registered Office

John Wiley & Sons, Inc., 111 River Street, Hoboken, NJ 07030, USA

Editorial Office

111 River Street, Hoboken, NJ 07030, USA

For details of our global editorial offices, customer services, and more information about Wiley products visit us at www.wiley.com.

Wiley also publishes its books in a variety of electronic formats and by print-on-demand. Some content that appears in standard print versions of this book may not be available in other formats.

Library of Congress Cataloging-in-Publication Data

Names: American Institute of Chemical Engineers. Center for Chemical Process Safety, author.

Title: Essential practices for creating, strengthening, and sustaining process safety culture / Center for Chemical Process Safety of the American Institute of Chemical Engineers.

Description: New York, NY : American Institute of Chemical Engineers, Inc. :
   John Wiley & Sons, Inc., 2018. I Includes bibliographical references. I
   Identifiers: LCCN 2018023924 (print) I LCCN 2018024176 (ebook) I ISBN 9781119515142 (Adobe PDF)
   I ISBN 9781119515173 (ePub) I ISBN 9781119010159 (hardcover)

Subjects: LCSH: Chemical engineering--Safety measures.

Classification: LCC TP150.S24 (ebook) I LCC TP150.S24 E87 2018 (print) I DDC
   660/.2804--dc23

LC record available at https://lccn.loc.gov/2018023924

Disclaimer

TABLE OF CONTENTS

SUPPLEMENTAL MATERIAL AVAILABLE ON THE WEB

Additional content referenced in this book as well as an electronic form of the culture assessment tool are available at www.aiche.org/ccps/publications/guidelines-culture

ACRONYMS AND ABBREVIATIONS

AI	Asset Integrity
AIChE	American Institute of Chemical Engineers
ALARA	As low as reasonably achievable
ALARP	As low as reasonably practicable
ANSI	American National Standards Institute
API	American Petroleum Institute
BBS	Behavior based safety
CBT	Computer-based training
CCC	Contra Costa County
CCPS	Center for Chemical Process Safety
DCS	Distributed control system
DIERS	Design Institute for Emergency Relief Systems of the American Institute of Chemical Engineers
EHS	Environmental, health, and safety
FMEA	Failure modes and effects analysis
HAZCOM	Hazard Communication (Standard – a U.S. regulation)
HAZOP	Hazard and Operability (Study)
HAZWOPER	Hazardous Waste Operations and Emergency Response (Standard – a U.S. regulation)
HIRA	Hazard Identification and Risk Analysis
HSE	Health and Safety Executive (United Kingdom)
IPL	Independent protection layer
ITPM	Inspection, testing, and preventive maintenance
ISA	International Society of Automation (formerly Instrument Society of America)
ISO	International Standards Organization, Industrial Safety Ordinance
ISD	Inherently safer design
LOPA	Layer of protection analysis
MI	Mechanical Integrity

MKOPSC	Mary Kay O'Connor Process Safety Center (Texas A&M University)
MOC	Management of Change
NEP	National Emphasis Program
NFPA	National Fire Protection Association
OE	Operational excellence
OSHA	Occupational Safety and Health Administration
PDCA	Plan-Do-Check-Act
PHA	Process hazard analysis
P&ID	Piping and instrument diagram
PSI	Process safety information
PSM	The USA OSHA Process Safety Management Regulation
PSMS	Process safety management system
PSSR	Pre-start-up safety review
QRA	Quantitative risk analysis
RAGAGEP	Recognized and generally accepted good engineering practice
RBI	Risk-based inspection
RBPS	Risk-based process safety
RC	Responsible Care®
RCA	Root cause analysis
RMP	Risk management program/risk management plan
RP	Recommended practice
SDS	Safety Data Sheet
SIL	Safety integrity level
SIS	Safety instrumented system
SOP	Standard operating procedure
SWP	Safe work practice
UKHSE	Health and Safety Executive (United Kingdom)
VPP	Voluntary protection program

GLOSSARY

CCPS has developed a standard glossary that defines many common terms in process safety. By reference the current CCPS Process Safety Glossary at the time of publication is incorporated into this book and can be found at http://www.aiche.org/ccps/resources/glossary. Additionally, there are some specific terms used in this book that are not currently included in the standard glossary. These terms are defined in the book as necessary when they are introduced.

ACKNOWLEDGEMENTS

The American Institute of Chemical Engineers (AIChE) and the Center for Chemical Process Safety (CCPS) thank the Process Safety Culture Subcommittee members and their CCPS member companies for their generous efforts and technical contributions to this book. CCPS also thanks the members of the CCPS Technical Steering Committee for their advice and support.

CCPS Process Safety Culture Subcommittee

The Chairs of the Process Safety Culture Subcommittee were Eric Freiburger of Praxair and Shakeel Kadri, then of Air Products and now CCPS Executive Director. The CCPS staff consultant was Bob Rosen. The Subcommittee members were:

Steve Arendt	ABS Consulting
Steve Beckel	Potash Corp.
Henry Brinker	Monsanto
Cho Nai Cheung	Contra Costa County
Gretel D'Amico	Pluspetrol
Michael Dossey	Contra Costa County
Walt Frank	CCPS Emeritus
Lou Higgins	Solvay Specialty Polymers
Tom Janicik	Covidien – Retired
Dave Jones	Chevron
Jack McCavit	CCPS Emeritus
Dan Miller	BASF (Retired)
Gilsa Pacheco Monteiro	Petrobras
Anne O'Neal	Chevron
Richard Piette	Suncor
Ravi Ramaswamy	Reliance Industries Ltd. (Retired)
Randy Sawyer	Contra Costa County
Karen Tancredi	Chevron

CCPS offers special thanks to subcommittee members Anne O'Neal, Karen Tancredi, Steve Arendt, Gretel D'Amico, Gilsa Pacheco Monteiro, and Dan Miller, who provided significant input during the production of this book.

CCPS acknowledge AcuTech Group, Inc. for preparing the original manuscript. The AcuTech Team, under the leadership of Project Manager Michael J. Hazzan, also included David M. Heller, Scott Berger, and Martin R. Rose. Lou Musante of Echo Strategies provided additional content under subcontract.

Scott Berger of Scott Berger and Associates LLC provided a structural edit of the manuscript, provided additional content, and typeset the final book.

Peer Reviewers

Before publication, all CCPS books undergo a thorough peer review. This book was no exception; many people offered thoughtful suggestions and comments.

Wendy Alexander	Nova Chemicals
Steve Barre	Huntsman Chemicals
Michael Broadribb	Baker Risk
Jonas Duarte	Chemtura
Warren Greenfield	Ashland Chemicals (retired)
Dennis Hendershot	CCPS Emeritus
John Herber	CCPS Emeritus
Jim Klein	ABS Consulting
Paul Leonard	Arkema
Bill McEnroe	Monroe Energy
Paul Nielson	Cheniere Energy
Lawrence Pearlman	Oliver Wyman
Dennis Rehkop	Tesoro
Susie Scott	Oliver Wyman

Mike Smith	Plains Midstream
Mark Trail	ExxonMobil - Retired
David Black	Baker Risk
Dan Wilczynski	Marathon Petroleum
Omer Wolff	Formosa Plastics USA

PREFACE

I have worked in different sectors of the chemicals and oil refining businesses since the 1960s. I began as a lab technician who worked shifts for a major chemical company in Northern Ireland, where I grew up. It was there that I experienced my first and only chemical process fatality during my working years in the chemical industry. I have very vivid memories of that tragedy to this day.

Later I went to work for a chemical company in the United States and I quickly realized it was vitally important to pay careful attention to preventing accidents as the chemicals we worked with included carbon monoxide, phosgene, chlorine, isocynanates and peroxides. In 1982, I served as the environmental manager in a chemical plant that had a catastrophic explosion. The details of that event and its aftermath are embedded deeply in my memory.

In 2002, I was appointed to the U.S. Chemical Safety Board (CSB) as a Board Member and later as Chairman. At the CSB we investigated failures in the chemical, oil refining and other industries – failures that resulted in loss of life, property damage and community outrage. Sadly, I saw many examples – fires, dust explosions, loss of containment, mechanical integrity failures.

In my early years in the chemical industry there was a strong focus on safety, but the emphasis was on the slips, trips and falls type of safety – avoiding injury to workers. Metrics were developed for first aid cases, reportable injuries and lost work day injuries. This was and still is a good practice and for the more progressive companies it made for a safer workplace environment. These companies were said to have a strong safety culture. Of course, process safety was still important, but not in an organized way. We knew the hazards of phosgene or dinitrotoluene and we took steps to mitigate those hazards. In the more enlightened companies, greater attention was paid to the chemical process hazards, but the culture in many companies was to equate overall safety with personnel safety, including some measures of process safety.

While the personnel safety record in the chemical and oil industry was better than general industry, unfortunately there continued to be major and well publicized fires and explosions in these industries. Tragedies such as the 1989 Phillips 66 explosion in Pasadena, Texas, the 1974 Nypro cyclohexane explosion in Flixborough, England, and the 1988 Shell refinery explosion in Norco, Louisiana. In response to these and other incidents, the U.S. Occupational Safety and Health Administration published its regulations on the process safety management of highly hazardous materials, commonly known as OSHA PSM. The 14 elements of PSM set an obligation for the safe operation of facilities with highly hazardous materials. The process industries have been required to comply with these regulations since 1992.

However, when the fourteen elements of PSM are examined there is an omission. That omission is the development and assessment of the process safety culture. I am very pleased that this absence has now been remedied by the publication of *Essential Practices for Creating, Strengthening and Sustaining Process Safety Culture* by the Center for Chemical Process Safety (CCPS) of the American Institute of Chemical Engineers. This excellent book fills a gap in the literature on process safety and guides companies and manufacturing facilities on the road to a strong process safety culture. It is the latest in a series of more than 100 high quality texts on process safety published by CCPS, many of which can be found on shelves in chemical plants and oil refineries around the world. Writing CCPS books requires the volunteer efforts of many experts from the chemical and oil industries. It is a time consuming but very satisfactory labor of love. I know because I have participated in the writing of a CCPS book.

This book offers several definitions of process safety culture. Even though there may be some disagreement about a definition of process safety culture, when you visit a facility you very quickly get a sense how important a positive process safety culture is to the facility. You will know it when you see it. From the first moment when you encounter a security guard or a receptionist to a tour of a control room you can quickly gauge the culture. Are process safety metrics displayed around the plant? Are operators communicating with each other in a professional manner? Is the senior manager well versed in the hazards of the operation?

As you read this book you will learn many aspects of how to develop a sound process safety culture. From my experience, a strong process safety culture must start with leadership. By leadership I mean everyone in a leadership position from the chairman of the board to the supervisor on the shop floor. They must set the example. It starts with leadership being aware of the hazards in their processes and putting in place the organization and expertise to control those hazards. Just as important, the senior leadership must communicate his or her concerns about the need for an effective process safety program. These concerns should be an ongoing part of senior leadership's communications with the organization. This is the way to ensure the establishment of a culture of process safety across the organization.

I commend CCPS on the publication of its latest book and I encourage readers to turn its lessons into actions in their day-to-day work of ensuring safety for employees, contractors and the surrounding community. As well as saving lives and preventing injuries it is vital for the financial success and reputation of the chemical process industries.

John S. Bresland
Shepherdstown, West Virginia

NOMENCLATURE

Culture: When used alone in this book, the term culture specifically means process safety culture, and the two terms are used interchangeably. When used to refer to other types of corporate culture, the specific type of culture will be specified, e.g. business culture.

Element Names: Process safety element names have been taken from CCPS *Guidelines for Risk Based Process Safety*. When alternative names are in common use, both the RBPS name and the common name are used, e.g., HIRA/PHA.

Operations: The full spectrum of tasks and activities involved in running a facility, including process operation, maintenance, engineering, construction, and purchasing.

Operator: An individual who runs the process from the control room and/or the field.

Process safety: A disciplined framework for managing the integrity of operating systems and processes handling hazardous substances by applying good design principles, engineering, and operating practices. It deals with the prevention and control of incidents that have the potential to release hazardous materials or energy. Such incidents can cause toxic effects, fire, or explosion and could ultimately result in serious injuries, property damage, lost production, and environmental impact.

Process safety management system (PSMS): A management system for implementing process safety. PSMSs include Risk Based Process Safety (RBPS) as defined by CCPS, the many PSMSs developed by companies to suit their specific requirements, PSMSs specified by regulations, and others.

References to process safety culture core principles: Throughout the book the names of the core principles of process safety culture are typeset in *italics*. Italics are also used when the context requires use of a different syntax, including the negative forms, such as "They allowed *deviance to be normalized*, leading to..."

Should vs. must and shall: The term *should*, used throughout the book, refers to actions or guidance that are recommended or presented as options, but not mandatory. The pursuit of process safety culture is very personal, and therefore a single approach cannot be mandated. The terms *must* and *shall*, commonly used in voluntary consensus standards and regulations, appear in this book only when quoting other sources. Quotes are offered only to provide perspective, and their use in this book does not mean that the authors consider the quoted text to be mandatory.

EXECUTIVE SUMMARY

Leading process safety practitioners have long recognized that the way leaders shape attitudes and behaviors can make the difference between success and failure in preventing catastrophic incidents. Investigations of incidents in the chemical, oil and gas sectors, as well as experience in the nuclear, and aerospace sectors have shown cultural failures rival management system failures as leading causes. Similarly, when long-term successes have been achieved, strong cultures of process safety excellence have been an integral factor.

This book provides current guidance on developing and improving process safety culture. It discusses how leaders can develop the commitment and imperative for process safety at the top, and then cascade that commitment throughout the organization. It shows how leaders can take the ultimate responsibility for process safety, and foster the core principles of process safety culture.

Of course, process safety culture does not exist in a vacuum relative to overall company culture. Changes to process safety culture may thus require changes in other aspects of the company culture, including, for example, operational excellence, human resources, and quality. This should not be viewed as a zero-sum game. Process safety may borrow key positive cultural attributes from other parts of the culture. Likewise, strengthening rocess safety culture may help strengthen other parts of the overall culture.

Leaders at any level of the organization will benefit from the guidance provided in this book. Senior executives will likely be drawn most to the first 3 chapters and the beginning of chapter 5, while the remainder of the book contains more detailed guidance useful at the implementation level. However, all readers will find useful information throughout the book.

After defining process safety culture, this book outlines 10 core principles of process safety culture:

- Establish an Imperative for Process Safety
- Provide Strong Leadership
- Foster Mutual Trust
- Ensure Open and Frank Communications
- Maintain a Sense of Vulnerability
- Understand and Act Upon Hazards/Risks
- Empower Individuals to Successfully Fulfill their Process Safety Responsibilities
- Defer to Expertise
- Combat the Normalization of Deviance
- Learn to Assess and Advance the Culture

The book then shows how these core principles strengthen process safety management systems (PSMSs), which implemented together can lead to success. The role of process safety culture in metrics, compensation, and other related activities is addressed. Lastly, the book discusses how to make process safety culture sustainable.

Appendices include more detailed descriptions of several concepts presented in the book, such as organizational culture, human behavior, and high reliability organizations, along with case histories useful for prompting culture discussions and a process safety culture assessment checklist.

The concepts discussed in this book began to be developed in the wake of the loss of the Space Shuttle Columbia. Members of CCPS toured the Columbia launchpad the day before launch as part of a learning-sharing session with NASA safety experts. This personal exposure to tragedy motivated Jones and Kadri (www.aiche.org/ccps, "Process Safety Culture Toolkit") to lead an effort to capture key culture lessons-learned from the Columbia investigation and apply them to the process industries.

Since that time, lessons continue to be learned about what makes process safety culture effective. This book attempts to distill the significant amount of published work, as well as the personal experience of CCPS member companies into actionable guidance.

Like other CCPS books, the guidance provided includes numerous options companies can choose from to suit their needs. While the book has been prepared with the similar care of a voluntary consensus standard, it is not a standard or a code, and has no legal or regulatory standing. And that is entirely appropriate to the mission of process safety culture – to create an imperative for process safety with felt leadership that comes from the heart, not forced by requirement.

1
INTRODUCTION

1.1 IMPORTANCE OF PROCESS SAFETY CULTURE

The 2014 FIFA World Cup semifinal between Germany and Brazil featured two of the most technically proficient teams to contest a match. Within a half-hour, however, the difference between the two emerged, as Germany scored five goals on a shell-shocked Brazil on the way to a 7-1 rout.

The difference? Neymar da Silva Santos, the captain, leader, and culture-setter of the Brazilian side, had suffered a fractured vertebra in the previous match, and could not even cheer his teammates on from the sidelines. With their culture-leader absent, Brazil failed to execute their usually formidable game plan and suffered a catastrophic loss.

Similarly, process safety cannot succeed without culture leadership. Investigation of numerous incidents in major hazard operations has clearly

> PSMS = Process Safety Management System

revealed culture deficiencies. The data show that without a healthy process safety culture, even the most well-intentioned, well-designed process safety management system (PSMS) will be ineffective. For example, Union Carbide was known as a process safety technology leader in the early 1980s. However, weak culture at its Bhopal facility allowed many "Normalization of Deviance" failures leading to the December 3, 1984 tragedy. Simply stated, a strong, positive process safety culture enables the

Essential Practices for Creating, Strengthening, and Sustaining Process Safety Culture, First Edition. CCPS. © 2018 AIChE. Published 2018 by John Wiley & Sons, Inc.

facility's PSMS to perform at its best. This gives the facility its best chance to prevent catastrophic fires, explosions, toxic releases, and major environmental damage.

Like all cultures, process safety culture starts with strong, committed, and consistent leadership. Just as commanding officers set the cultures of their troops, senior leaders of facilities and companies set the process safety culture of their organizations. Senior leaders set the underlying tone for how an organization functions and motivates the individuals within the organization to maximize the impact of their collective talent (Ref 1.1).

Without leadership's direct, continuing, and strong participation in setting process safety culture, the culture will suffer gaps in one or more of the ten cultural principals (see chapter 2). This leadership should cascade through the organization, with each leader helping their subordinates, peers, and managers maintain focus on achieving the desired culture.

Leadership of culture should survive economic downturns and keep pace with upturns and technology changes. Culture leadership should persist through acquisitions and divestitures. Perhaps hardest of all, it should survive changes of personnel. Altogether, leadership should be committed to establishing and maintaining a sound process safety culture and should establish the proper philosophical tone for the culture. This tone should emphasize the true importance of process safety and the faithful execution of the PSMS. The importance of strong leadership will be further discussed in section 1.4 and in Chapter 3.

1.2 DEFINITION OF PROCESS SAFETY CULTURE

Many experts have defined culture as what people do when their boss is not around. A group of people with a common purpose (e.g., co-workers, teammates, and families) develops a set of beliefs, customs, and behaviors that become embedded in how the group thinks and works. With continued practice, these beliefs and behaviors become reinforced and integrated into the group's value system (Refs. 1.2, 1.3). As time goes on, the group's actions reflect common and deeply held values. The group expects newcomers to adopt or "buy into" these values to become accepted into the group.

Unfortunately, negative cultures can also exist, where common values result in attitudes and actions with negative consequences. In such cultures, peer pressure can reinforce negative behaviors. This may happen for example, if a new co-worker berated for following the approved procedure instead of the common but unsafe shortcut.

The International Nuclear Safety Advisory Group (INSAG) of the International Atomic Energy Agency (IAEA) made one of the first definitions of safety culture in the investigation of the aftermath of the Chernobyl accident in 1986 (Ref 1.4).

"Safety Culture is that assembly of characteristics and attitudes in organizations and individuals which establishes that, as an overriding priority, nuclear plant safety issues receive the attention warranted by their significance."

The preceding definition describes the result of the culture, but not the culture itself (Ref 1.5). In the wake of the *Challenger* and *Columbia* disasters, NASA (Ref 1.6, 1.7) began to recognize that key personnel defined organizational culture, and that change in personnel can lead to negative culture change:

"Organizational culture refers to the basic values, norms, beliefs, and practices that characterize the functioning of a particular institution. At the most basic level, organizational culture defines the assumptions that employees make as they carry out their work; it defines "the way we do things here." An organization's culture is a powerful force that persists through reorganizations and the departure of key personnel."

Describing groundbreaking CCPS work in 2005, Jones and Kadri (Ref 1.8) adapted these published definitions to process safety and recognized the link of culture to management:

"For process safety management purposes, we propose the following definition for process safety culture: The combination of group values and behaviors that determine **the way process safety is managed**.*" (emphasis added)*

In the wake of its investigation of a refinery explosion in Texas City, TX, USA, the US Chemical Safety Board (CSB) leveraged the CCPS work Jones and Kadri described (Ref 1.9). CSB recommended that the company conduct an independent assessment of process safety culture at their five U.S. Refineries and at the Corporate level. The resulting Baker Panel report (Ref 1.10 identified numerous culture gaps and improvement opportunities. They then went on to say, "We are under no illusion that deficiencies in process safety culture, management, or corporate oversight are limited to the company." This statement proved to motivate many process safety culture improvements in refining and chemical companies globally.

Additional study led CCPS to define process safety culture based on the critical role of leadership and management. CCPS's Vision 20/20 (Ref 1.11) CCPS stated that a committed culture consists of:

1. Felt leadership from senior executives. Felt leadership means more than a periodic mention of process safety in speeches and town hall meetings. It means that executives feel a deep personal commitment and remain personally involved in process safety activities.

2. Maintaining a sense of vulnerability.

3. Operational discipline, the performance of all tasks correctly every time.

This sums up several definitions of culture from other sources as it applies to environmental, health, or safety programs and issues:

- **(Ref 1.12):** "Safety and health are (or have become) part of the company culture—and frequently part of the management system. 'Culture' is traditionally defined as 'a shared set of beliefs, norms, and practices, documented and communicated through a common language.' The key word here is 'shared.' Companies have found that if safety and health values are not consistently (and constantly) shared at all levels of management and among all employees, any gains that result from declaring safety and health excellence a 'priority' are likely to be short-lived."
- **(Ref 1.13):** "The attitudes, beliefs and perceptions shared by natural groups as defining norms and values, which determine how they act and react in relation to risks and risk control systems."
- **Canadian National Energy Board (Ref 1.14):** "Safety culture means 'the attitudes, values, norms and beliefs, which a particular group of people shares with respect to risk and safety'."
- **UK Health and Safety Executive (Ref 1.15):** "The safety culture of an organization is the product of individual and group values, attitudes, perceptions, competencies and patterns of behaviour that determine the commitment to, and the style and proficiency of, an organization's health and safety management."

These definitions share common themes and terms. For something to become embedded in the culture of an organization of group, it is believed by its members. The belief becomes a common or shared belief, a value, or a norm. These norms result in certain repeated actions or behaviors.

The shared beliefs and values may create a culture that is either positive or negative, either strong or weak. A strong positive process safety culture would generally exhibit norms such as:

- Always doing the right thing even when nobody is watching or listening,
- Not tolerating deviance from approved policies, procedures, or practices,
- Maintaining a healthy respect for the risks inherent to the processes, even when the likelihood of serious consequences is very low; and
- Performing actions safely, or not performing them at all.

Conversely, a negative or weak culture would generally exhibit norms such as:

- Tolerating deviance from approved policies, procedures, or practices,
- Allowing such deviance to become regular occurrences,
- Exhibiting complacency regarding the operation's process risks; or
- Allowing short-cuts to occur to get something done more quickly or more cheaply.

The CCPS Culture Subcommittee distilled the published definitions listed above, along with their personal ongoing experience in building and strengthening process safety culture. For purposes of this book, a sound or strong positive process safety culture is:

> *The pattern of shared written and unwritten attitudes and behavioral norms that positively influence how a facility or company collectively supports the successful execution and improvement of its Process Safety Management System (PSMS), resulting in preventing process safety incidents.*

From this starting point, Chapter 2 will describe core principles of process safety culture. Chapter 3 will discuss the leadership dimensions of culture. Chapter 4 will address culture from the standpoint of organizational dynamics, human behavior, compensation, ethics, external influences (e.g. contractors, vendors, public sector), and metrics. Chapter 5 will discuss the ways in which culture can directly impact each element of CCPS's Risk Based Process Safety (RBPS) PSMS. Chapter 6 will provide a guide for getting started establishing a strong culture or improving culture. Chapter 7 will address how to achieve a sustainable culture. The appendices provide additional background on culture, case histories that may be useful in discussing culture issues, and a culture assessment protocol. Taken together, the concepts discussed in these chapters provide the concepts and guidance to make these concepts a reality in an organization.

This book does not discuss regulations, but instead comes from the point of view that a strong positive culture adequately addresses process hazards, whether regulated or not. This represents the first concept of a strong, positive process safety culture: the organization's leadership and all personnel believe in the necessity of process safety and commit to it, even in the absence of regulatory requirements.

Some people have expressed the belief that safety culture cannot change. They consider core principles, company values and principles, and how the company behaves. They then conclude that good cultures will stay good, while poor cultures cannot improve. Mathis disagrees, suggesting that those who claim culture is static may be resisting the culture change (Ref 1.5).

From a sociology point of view, cultures of all kinds develop via social conditioning. With the right conditioning, applied patiently over time, leaders can build strong positive cultures. Typically, this requires patience and persistence. It can take some time to build workers' trust and to convince them that the intended culture change is not a temporary fad.

Conversely, negative conditioning can occur. Since trust can be lost much faster than it can be gained, even momentary lapses in process safety leadership can lead to rapid degradation in the culture.

Clearly then, process safety has an inherent capability to improve – and to degrade, and no single culture resides in the DNA of the organization. This makes it essential to have the patience to improve of process safety culture over time, and then maintain focus on culture over time to maintain consistent good performance.

While this book addresses process safety culture, the concepts of process safety culture are not unique to process safety. Good concepts may be leveraged from the overall company culture or various subcultures (e.g. the company's innovation, sales, financial, EHS, and other cultures). At the same time, if any of these company cultures contain values contrary to a good process safety culture, leaders need to recognize this and find a way to keep those values out of the process safety culture effort.

The company and facility's country and regional cultures should also be considered. These can make a culture effort either easier or more difficult. Diversity of the organization's personnel can inject a wider range of external cultures that could impact a culture effort. Essentially, everyone within a diverse group of employees will have to make unique kinds of culture changes to arrive at the desired common process safety culture. Diversity may also inject different languages or different ways that things are expressed into a given facility, and should be accommodated in the communications between personnel. Leaders of culture change need to consider these factors.

Diversity plays an even more significant role when a company strives to establish a global process safety culture. Each facility has a distinct culture. Any facility may have positive cultural aspects that will help the process safety culture transformation effort, just as they may have negative cultural aspects. The combination of positives and negatives may help some facilities transform culture more quickly, while other facilities may require considerable time and effort to transform. For this reason, companies may choose to take approaches to culture that leverage regional cultural strengths.

Additionally, process safety culture may not be completely uniform across a given company or site. While certain core values should exist, subcultures can and often will exist within an organization. These subcultures can vary by facility (one site vs another), workgroup (Inspection vs. Instrument/Electrical within Maintenance), occupation or discipline (e.g., Engineering vs. Operations), age, work shift, prior accident involvement, contractor vs. full-time employee, groups within an organization subject to different working conditions, and grade (management vs. non-management) (Ref 1.15). The difference in culture between daytime and nighttime shifts and between management and non-management can sometimes be significant.

Such groups will tend to view safety through the lens of their own subcultures, rather than sharing an overall view of safety. The presence of subcultures within an organization can lead to misunderstandings and ultimately conflict between groups. The investigation into the Piper Alpha disaster (Ref 1.16) highlighted the lack of communication between the day and night shifts, despite there being shift hand-over and permit to work systems. However, subcultures can also be a positive influence on safety, by bringing different perspectives and a diversity of views to safety problems.

The culture of outside groups that interact with a facility can also influence its process safety culture. These outsiders include contractors, regulators, law enforcement, emergency responders, media, unions, corporate staff, boards of directors, interest groups, community groups, individual members of the public, among others. For example, the UK HSE (Ref 1.15) found that contract workers on offshore platforms had "markedly inferior" benefits and working conditions than employees. Contract workers did not receive holiday or sick pay, and they did the most dangerous and physical work. Their working conditions distanced them from the company's culture. Unlike employees, contractors viewed safety as subordinate to production. Unsurprisingly, contractors experienced more accidents than company employees.

Multi-cultural influences work in both directions. The cultures of the outsiders will affect the culture of the facility, and the culture of the facility will affect the culture of the outsiders. Therefore, a facility should communicate and partner with the outsiders to advance their process safety related cultures as they advance their own. Each of these outside groups has its own agenda and interests, and sometimes these interests will not be in accord with those of the company.

The way in which different parties communicate with each other, build trust, and resolve conflict will determine how much friction results. Chapter 5 addresses the external influences on culture and how it relates to the facility's and company's PSMS.

The CCPS RBPS book (Ref 1.3) includes process safety culture as a distinct PSMS element and defines some tangible actions. However, many of the recommended actions to establish a strong process safety culture are intangible. A healthy process safety culture cannot be successfully established by edict or by cookbook. It requires convincing employees and the abovementioned outsiders that a healthy process safety culture benefits them. The rationale should be carefully and fully explained, supported with empathy, and led by example in implementation. The tangible actions described in the RBPS book, such as formally defining process safety goals and objectives, formally defining responsibilities, accountabilities, and training requirements will be described in more detail in Chapters 4, 5, and 6.

1.3 WARNING SIGNS OF POOR PROCESS SAFETY CULTURE

In its book *Recognizing Catastrophic Incident Warning Signs in the Process Industries* (Ref 1.17), CCPS described typical warning signs that have been observed in facilities well in advance of incidents. CCPS argues that recognizing the presence of these warning signs allows companies to correct deficiencies early before they escalate into incidents.

Table 1.1 lists warning signs related to Process Safety Culture, organized by core culture principles that will be presented in chapter 2. Table 1.1 is organized in the form of a checklist that readers can use to find quick culture improvement opportunities before launching a formal culture improvement effort. This checklist may also be used to identify the core culture principles on which to focus initial work. It can also be useful to stimulate discussions about culture in leadership team meetings.

Table 1.1 Checklist – Warning Signs of Gaps in Process Safety Culture

Warning Sign	Found?	Action
Establish an Imperative for Safety		
• Widespread confusion between occupational safety and process safety		
• Process safety budget reduced		
• Leadership behavior implies that public reputation is more important than process safety		
Provide Strong Leadership		
• Frequent organizational changes		
• Conflict between production goals and safety goals		
• Employee opinion surveys give negative feedback		

Table 1.1 (Continued)

Provide Strong Leadership (Cont'd)		
• A high absenteeism rate		
• Frequent changes in ownership		
Foster Mutual Trust		
• Strained communications between management and workers		
• A lack of trust in field supervision		
• Favoritism exists in the organization		
• An employee turnover issue exists		
Ensure Open and Frank Communications		
• Negative external complaints		
• Inappropriate supervisory behavior		
Maintain a Sense of Vulnerability		
• Leaders obviously value activity-based behavior over outcome-based behavior (sometimes referred to as "check-the-box" activities or a "checklist approach")		
Understand and Act Upon Hazards and Risks		
• Signs of worker fatigue		
• Overdue process safety action items		
• Workers not aware of or not committed to standards		
Empower Individuals to Successfully Fulfill their Safety Responsibilities		
• Job roles and responsibilities not well defined, confusing, or unclear		
• Conflicting job priorities		
• Frequent changes in priorities		

Table 1.1 (Continued)

Empower Individuals … (Cont'd)		
• Supervisors and leaders not formally prepared for management roles		
• A poorly defined chain of command		
Defer to Expertise		
• Conflict between workers and management concerning working conditions		
• A perception that management does not listen		
Combat the Normalization of Deviance		
• Operating outside the safe operating envelope is accepted		
• Varying shift team operating practices and protocols		
Learn to Assess and Advance the Culture		
• Slow management response to process safety concerns		
• Everyone is too busy (helps foster "check-the-box" thinking		

1.4 LEADERSHIP AND MANAGEMENT ROLES AND RESPONSIBILITIES

The responsibility for executing the PSMS and maintaining the process safety culture belongs to those persons who manage the company and its facilities. Process safety culture specifically is created through leadership.

Senior leaders have high-level and strategic oversight for both the management system and the culture. They delegate program and element responsibilities through the organization, and they create the culture. Leadership for process safety needs to start at the top of the organization, and then carry through all levels, to the plant floor. Strong leadership stewarding each function is needed to direct resources to the most critical risks and opportunities, clarify expectations, listen and learn, create passion, and provide clear, consistent messages. As discussed in Chapter 2, Leadership is a core principle of process safety. The role of management and leadership will be discussed in more detail in Chapter 3.

In a complex business with a high-risk profile that suffers from lack of leadership, cultural gaps will appear and can lead to process safety performance gaps. These in turn can lead to catastrophic incidents. Therefore, leaders in the chemical, oil and gas, and related industries have no role more important than stewarding the PSMS and process safety culture in their organizations.

What causes process safety cultures to fail? Roughton and Mercurio (Ref 1.18) state that in many cases these failures occur due to management style. Their research identified two primary types of management styles: authoritarian management and participative management.

Authoritarian managers stress productivity and often believe that people inherently avoid work. They operate by command and control, which may get tasks done. But they fail to motivate people because they do not fulfill basic human social and ego needs. Furthermore, this management style limits ingenuity, creativity, and problem-solving to only a few individuals, only partially utilizing the intellectual potential of the workplace.

Participative managers recognize that people can be positively motivated by the satisfaction of doing their job well. Accordingly, direct control and punishment can be successfully replaced by self-direction. Workers committed to their work seek responsibility rather than avoid it. They then exercise their capacity for imagination, ingenuity, and creativity, often to the benefit of both safety and productivity.

Roughton quotes an old military adage that summarizes the importance of leadership to successful management: "There are no bad troops, only bad officers." Clearly, leadership is integral to establishing a sound culture.

1.5 ORGANIZATIONAL CULTURE, PROCESS SAFETY CULTURE, AND BUSINESS SUCCESS

CCPS (Ref 1.3) describes four foundational blocks of a PSMS:

- Commit to Process Safety,
- Understand Hazards and Risks,
- Manage Risks; and
- Learn from Experience.

This links to the familiar Plan-Do-Check-Act (PDCA) cycle as shown in Figure 1.1, with leadership providing the culture that drives success.

Figure 1.1
Linkage of Culture and Leadership to a PSMS

Strong, positive leadership in process safety will establish the groundwork for a sound process safety culture. That sound culture will then lead to a robust PSMS that in turn drives improved and sustained process safety performance.

Since process safety follows the PDCA approach used in other operational and business systems, improving process safety culture will also likely lead to improvements in other cultures, such as EHS, Quality, and technology, and therefore lead to stronger business performance (see section 1.5 and Appendix A).

Likewise, process safety culture of an organization does not exist in a vacuum. Instead, it inextricably links to the organization's overall culture, including other subcultures such as business practices, overall EHS, quality, and even the culture of stakeholders that interface with the organization (e.g., neighbors, customers, etc.), and others.

In the ideal case, a strong positive process safety culture mates with other strong positive cultures to build an overall strong positive corporate culture, as discussed by Musante (Ref 1.19). This kind of strong and integrated culture is sometimes referred to as Operational Excellence. The Musante reference, titled *Doing Well by Doing Good: Sustainable Financial Performance Through Global Culture Leadership and Operational Excellence,* is reproduced with permission as Appendix A.

A weak or negative process safety culture may be coupled with, for example, a strong business culture. This may provide financial success and avoid process incidents for some time, but ultimately a major process safety incident can cause it to fail catastrophically. However, strong business culture can be leveraged to build a strong process safety culture. Likewise, when both process safety and business cultures are relatively weak, first strengthening the process safety culture can be a stepping stone to building an overall positive business culture.

1.6 CORPORATE CLIMATE AND CHEMISTRY

If process safety culture underpins everything in a PSMS, then what underpins the culture? What conditions either support or inhibit the development, maintenance, and sustainability of the process safety culture? Mathis and Galloway (Ref 1.5). identify seven milestones on the safety culture improvement journey. Two of those milestones are climate and chemistry.

Corporate Climate refers to the conditions within an organization as viewed by its employees. In the case of process safety, management creates an organization's climate through four components: Commitment, Caring, Cooperation, and Coaching. Two organizations may have a common set of activities, from which an external viewer might infer the same culture. However, the cultures may be very different. For example, soldiers in combat and participants in a survival reality TV show may share some common tasks, i.e., surviving in harsh outdoor conditions, but the climate or environment for these two situations are totally different and therefore the cultures will be very different.

Corporate Chemistry refers to the structure of the culture. Like the elements that make up a molecule or the elements in the soil that nurtures the growth of plants, safety culture is built around the elements of Passion, Focus, Expectations, Proactive accountability, Reinforcement, Vulnerability, Communication, Measurement, and Trust. (Ref 1.5).

In developing the culture principles presented in this book, CCPS considered both climate and chemistry. Some culture principles addressed both climate and chemistry, as shown in figure 1.2, facing.

1.7 SUMMARY

Process safety culture has been recognized as a contributing factor in many significant incidents that have occurred in the processing industries in recent years. Process safety culture in any

Figure 1.2 Mapping of CCPS Culture Principles to Corporate Climate and Chemistry

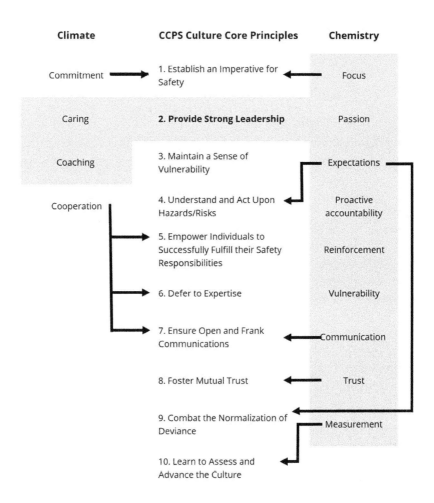

facility forms the foundation of the PSMS, regardless of what is written. The quality of demonstrated leadership directly affects the strength and quality of the process safety culture, and the quality and health of the process safety program itself. While managers of PSMS's clearly serve as process safety leaders, all managers and executives in enterprises that manage major process hazards can and indeed should demonstrate process safety leadership and help set a strong, positive culture.

CCPS defines process safety culture as:

"The pattern of shared written and unwritten attitudes and behavioral norms that positively influence how a facility or company collectively supports the development of and successful execution of the management systems that comprise its process safety management system, resulting in the prevention of process safety incidents."

Other definitions related to safety culture and particularly process safety culture can be found in the literature. There is no single definition of safety or process safety culture. Numerous definitions have been presented in the literature in recent year. The CCPS definition embodies all the lessons learned in the literature to produce a definition serving the major hazard industries ranging from upstream oil and gas, through refining, chemicals and pharmaceuticals to manufacturers who handle chemicals and practice chemistry in other industries.

Process safety culture and the organizational culture that it fits into are strongly linked. The process safety/safety culture of an organization cannot exist in a vacuum. Any problems or issues in the organizational culture will also show up in the process safety culture, and elsewhere in the organization as well. Likewise, efforts to improve process safety culture can spill over to positively impact the overall culture of the organization. Organizations that have a strong overall culture and strong process safety culture have been shown to consistently have better financial performance. There is a strong business case for strengthening and sustaining process safety culture.

1.8 REFERENCES

1.1   Sielski, M., The Philadelphia Inquirer, October 25, 2014.

1.2   Schein, E.H., *Organizational Culture and Leadership*, 3rd Ed., Jossey-Bass, 2004.

1.3   CCPS, *Guidelines for Risk Based Process Safety*, American Institute of Chemical Engineers, New York, 2007.

1.4   International Atomic Energy Agency (IAEA), *Safety Series No. 75 – INSAG-4, Safety Culture*, 1991.

1.5   Mathis, T., Galloway, S., *STEPS to Safety Culture Excellence*^{SM}, Wiley, 2013.

1.6   National Aeronautics and Space Administration, *Columbia Accident Investigation Board Report*, Washington, DC, August 2003.

1.7   Rogers, W.P. et al., *Report of the Presidential Commission on the Space Shuttle Challenger Accident, Washington, DC*, June 6, 1986.

1.8   Jones, D., Kadri, S., *Nurturing a Strong Process Safety Culture*, Process Safety Progress, Vol. 25, No. 1, American Institute of Chemical Engineers, 2006.

1.9   CCPS, Process Safety Culture Tool Kit, American Institute of Chemical Engineers, New York, 2004.

1.10   Baker, J.A. et al., *The Report of BP U.S. Refineries Independent Safety Review Panel*, January 2007.

1.11   McCavit, J, Berger, S., Grounds, C., Nara, L., *A Call to Action - Next Steps for Vision 20/20*, CCPS 10th Global Congress on Process Safety, New Orleans, 2014.

1.12   Whiting, M. and Bennett, C., *The Conference Board, Driving Toward '0': Best Practices in Corporate Safety and Health*, Research Report No. R-1334-03-RR, 2003.

1.13   Hale, A.R., *Culture's Confusions*, Safety Science, Vol. 34, No. 1-3 (2000).

1.14   Canadian National Energy Board, *Advancing Safety in The Oil and Gas Industry Statement on Safety Culture*, from Mearns, K., Flin, R.,

Gordon, R. & Fleming, M. (1998), *Measuring safety culture in the offshore oil industry*, Work and Stress, 1998.

1.15 United Kingdom Health and Safety Executive (HSE) Health & Safety Laboratory, *Safety Culture: A review of the literature,* HSL/2002/25, 2002.

1.16 HM Stationery Office, *The Public Inquiry into the Piper Alpha Disaster*, Cullen, The Honourable Lord, 1990.

1.17 CCPS, *Recognizing Catastrophic Incident Warning Signs in the Process Industries,* American Institute of Chemical Engineers, New York, 2012.

1.18 Roughton, J, Mercurio, J, *Developing an Effective Safety Culture: A Leadership Approach*, Butterworth-Heineman, 2002.

1.19 Musante, L. et al., *Doing Well by Doing Good: Sustainable Financial Performance Through Global Culture Leadership and Operational Excellence,* Echo Strategies, October 2014.

2
PROCESS SAFETY CULTURE CORE PRINCIPLES

In its early work on process safety culture (Ref 2.1), CCPS identified 6 themes of process safety culture:

- Maintain Sense of Vulnerability
- Combat Normalization of Deviance
- Establish an Imperative for Safety
- Perform Valid/Timely Hazard/Risk Assessments
- Ensure Open and Frank Communications
- Learn and Advance the Culture

In preparing this book, the culture-related causes of major incidents in various industries have been studied. As a result, these 6 themes have been expanded and developed into 10 core principles. These 10 core principles are summarized in figure 2.1 and discussed in detail in this chapter. Four new principles have been added, and the original theme regarding hazard/risk assessment has been expanded to clarify that action should result from these assessments. This chapter describes these core principles, where they come from, their fundamental basis, and some key indicators.

Essential Practices for Creating, Strengthening, and Sustaining Process Safety Culture, First Edition. CCPS. ©2018 AIChE. Published 2018 by John Wiley & Sons, Inc.

Figure 2.1 Overview of the Core Principles of Process Safety Culture

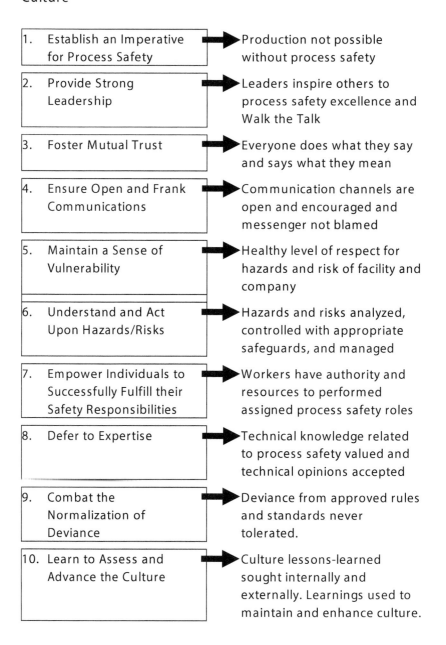

The 10 core principles have some overlap. Readers may note, for example, the similarity of Core Principle 7 (Empowering individuals) and Core Principle 8 (Deferring to their expertise). Nonetheless, the activities associated with these related elements are different, and that differentiation helps provide clarity in the presentation of these Guidelines.

The order of the Core Principles shows the dependency of each Core Principle on others. Ultimately, to successfully implement the later principles, a solid foundation should be built upon the earlier principles. Indeed, company and site leadership should make a conscious business commitment to process safety and internalize it personally before making significant efforts in the other Principles. With these in place, leaders then have the possibility to build trust and communication, and start implementing the remaining Principles.

2.1 ESTABLISH AN IMPERATIVE FOR PROCESS SAFETY

Illiopolis, Illinois, USA, April 23, 2004

An explosion and fire of vinyl chloride monomer killed five workers and severely injured three at a polyvinyl chloride (PVC) manufacturing facility (Ref 2.2). A worker overrode the interlock to prevent opening the bottom valve on a pressurized reactor. As a result, hot vinyl chloride monomer spewed into the building, ignited, and exploded. The explosion destroyed most of the plant. Smoke from the fire drifted over the local community. As a precaution, local authorities evacuated the community for two days.

To override the interlock, the worker used a dedicated, labeled "Emergency Air Hose." This hose had a specific emergency use, requiring authorization from a senior manager designated to approve such variances. However, the plant had recently been acquired. In the restructuring that followed, the approving manager's position was eliminated without reassigning that key responsibility. As a result, operator began self-approving the bypass. Through its actions, the acquiring company failed to **Establish an Imperative for Process Safety**. Failure to *Establish an Imperative for Process Safety* also existed in the failure to follow the emergency evacuation procedure, contributing to the severity of the incident.

Weaknesses in other culture core principles contributed to the incident. Without the approver, the operators began regularly using the hose against procedure to force open stuck valves. This shows a failure to *Combat the Normalization of Deviance*. An incident with a similar cause had occurred in the company just a few months earlier. However, the company failed to *Learn to Assess and Advance the Culture*.

Most managers and executives will extoll the importance of safety. But this alone will not establish an imperative for process safety. Without frequent, serious demonstration of the value of process safety, everyone throughout the organization will recognize the hollowness of the words and slogans, and fail to be convinced.

Leaders demonstrate the imperative for process safety when they demonstrate that production depends on process safety. In doing so, a facility should have to prove that it can operate safely. Likewise, workers, supervisors, and process safety personnel should never be put in the untenable position of having to prove that an operation is unsafe. The imperative for process safety is tested when leadership emphasizes meeting work demands, schedule, or budget over process safety. Sometimes the emphasis can be subtle, such as in the way performance measures weighted.

The work to establish an imperative for process safety parallels any other culture change a company would wish to make. A clear organizational vision and mission should be established, ideally with input from the front line. The vision and mission should be exciting, and be urgent enough to create excitement and motivation for the employees. At the same time, it should be achievable, so that it is not shrugged off as impossible. Finally, the vision and mission should have a long-term outlook, to assure everyone that the new process safety culture is here to stay. (Ref 2.3)

Signs that an imperative for process safety have been established include:

- The organization is very attentive to safety and process safety. The organization can anticipate areas of potential failure and maintain resilient processes and systems that can survive upset and return to normal operation despite challenges. (Ref 2.4)
- Operational aspects of process safety are integrated into operations, but corporate oversight for process safety is maintains independence. This helps align process safety with operations, while helping avoid conflict of interest. Smaller organizations with limited staff with multiple functions should think this through carefully (see also section 4.3 ethics).
- Process safety resources survive budget cuts during downturns. This does not mean that process safety budgets are never cut. However, the process safety competency required to operate safely should not be compromised.
- Management praises process safety as a value to the company. The process safety organization has strengthened due to its performance and has gained influence in the decision-making process.

- Those responsible for process safety are fully qualified to do the job.
- Process safety staff is not placed in the untenable position of having to prove that an operation is unsafe. Those desiring or advocating certain operations or conditions should be required to prove that those operations or conditions are safe.
- Process safety metrics and audits are used to guide improvement. They are not treated as adversarial or punitive activities.
- The imperative for process safety extends equally to contractors, labor unions, headquarters staff, and outside members of the Board of Directors. To the degree possible, the imperative also extends to community members, public interest groups, and regulators (see also section 4.4).

The Baker Panel (Ref 2.5) noted that commercial considerations, including cost control and production, play a role in defining the safety culture of an organization. All organizations that produce goods and services not only face limitations on human and financial resources, but also must effectively manage the tension that exists between the operational demands relating to production and the demands relating to safety. Reason (Ref 2.6) summarized this natural tension:

"It is clear from in-depth accident analyses that some of the most powerful pushes towards local [culture] traps come from an unsatisfactory resolution of the inevitable conflict that exists (at least in the short-term) between the goals of safety and production. The cultural accommodation between the pursuit of these goals must achieve a delicate balance. On the one hand, we have to face the fact that no organization is just in the business of being safe. Every company must obey both the 'ALARP' principle (keep the risks as low as reasonably practicable) and the 'ASSIB' principle (and still stay in business)."

The balance an organization strikes between safety and production considerations, and how it organizes itself to accomplish this balance, serves in part to define the organization's safety culture. An organization should recognize that a wide variety of stakeholders, including owners/shareholders, managers, workers, and the public at large, have an interest in its safety culture. Moreover, an organization with a strong safety culture does not lose sight of the fact that the stakeholders with the most to lose—their lives— are the workers in hazardous operating units and members of the public living nearby.

When the imperative for process safety has been established, the organization and all its employees and contractors truly believe in the value of a robust PSMS. Everyone in the organization knows their role in protecting the facility, their co-workers, the public, and the environment from potential catastrophic incidents. The leaders' – and everyone's – words are supported with resources and actions.

2.2 PROVIDE STRONG LEADERSHIP

Gulf of Mexico, Offshore of Louisiana, USA, April 20, 2010

Eleven workers died and seventeen suffered serious injuries when a deep-water oil well blew-out (Ref 2.7). Hydrocarbon liquids and gas blew back and released on the platform. The resulting explosion and fire sunk the platform. The well kick also disabled the backflow preventer at the wellhead, allowing oil and gas to spew uncontrollably into the Gulf. Five million barrels of oil spilled before the well could be capped nearly 3 months later. Ironically, the incident occurred while management visited the rig to celebrate good occupational safety performance.

Many process safety culture gaps, along with technical and management system deficiencies contributed to this incident.

The *Imperative for Process Safety* did not measure up to the imperative for occupational safety, a theme that will be often repeated in this book. However, all the other cultural gaps trace back to a failure to *Provide Strong Leadership* for process safety.

Workers attempting to control the well did not have the physical capability to perform the tasks required. This speaks to a weak *Sense of Vulnerability* as well as insufficient *Understanding and Action on Hazards and Risks*. These two gaps also influenced the over-reliance on the un-tested Blowout Preventer as a single safeguard. Investigators also noted numerous short-cuts from intended procedures that had become routine (*Normalization of Deviance*).

This incident also highlighted issues in *Open and Frank Communication* between the owner-operator and the contractor, leading to poorly defined assignment of process safety roles.

In this book, the terms process safety leaders and leadership apply to the senior executives of the company and the line organization. Clearly, process safety professionals are also expected to provide leadership as well. However, time and again, experience shows how driving process safety excellence through the line organization achieves the best results.

The Organization of Economic Cooperation and Development (Ref 2.8) highlighted the vital role senior leaders play in a process safety program:

"Strong leadership is vital, because it is central to the culture of an organisation, and it is the culture which influences employee behaviour and safety. Process safety tasks may be delegated, but responsibility and accountability will always remain with the senior leaders, so it is essential that they promote an environment which encourages safe behaviour.

"Creating a culture where all employees expect the unexpected and strive for error-free work is absolutely essential for success in process safety. This kind of culture is possible only through demonstrated leadership at all levels of the organisation."

Showing strong leadership in process safety means convincing direct reports, peers, and even superiors of the right process safety thoughts and actions. To be truly effective, leaders earn respect, and inspire others to support and take the proper process safety actions. Leaders set a vision, mission, and goals for process safety and show passion for achieving these, along with setting priorities and providing resources. Process safety leaders establish the environment for succeeding in the other culture core principles, especially Open and Frank Communication, Empowerment, and Trust. This kind of leadership must be taken willingly and made personal. It does not come automatically with a position, an impressive title and assigned authority. Process safety leaders win both hearts and minds.

In CCPS Vison 20/20 (Ref 2.9), CCPS defines the concept of felt leadership. Felt leadership means that the executives and other leaders personally involve themselves in process safety activities. Employees know the executives care about process safety because of what they see and feel executives doing, not just by they hear them saying. Felt leadership is leading by passionate example.

What leaders do and how they appear to feel about it can have a powerful influence on their followers. This can be even more clearly seen in the violation of this principle. When leaders say that process safety is their top goal one thing, and then fail to support it, their followers conclude that process safety goal was not important after all.

Leaders should show tangible commitment and influence the same commitment in their subordinates. Tangible commitment comes when leaders, and the leaders who report to them, provide the needed human and financial resources, hold staff accountable for their goals, and engage personally regarding process safety with employees at all levels of the organization.

Chapter 3 of this book further discusses the importance of leadership on the process safety culture, including the difference between management and leadership, the technical competence of leadership, and other relevant topics.

2.3 FOSTER MUTUAL TRUST

Texas City, Texas, USA, March 23, 2005

A vapor cloud explosion during startup following a turnaround in a refinery's isomerization unit resulted in 15 fatalities and over 170 injuries. During start-up, operators introduced feed and began heating without controlling the column level. Hydrocarbon filled the column and boiled over into the atmospheric blowdown stack that sprayed hydrocarbon into the air. Gaps in nearly every RBPS element contributed to the disaster. (Ref 2.10).

This incident proved to be a watershed event in recognizing culture as a key aspect of process safety. Following the incident, a panel of experts chaired by former U.S. Secretary of State James A. Baker, III convened to provide an independent assessment of the process safety culture the company's U.S. refineries. The Baker Panel report (Ref 2.5), published in 2007, identified many ways that culture contributed to the event. Among its many findings, the Panel noted a failure to *Foster Mutual Trust* between the refinery, which had operated under another company until its acquisition, and the headquarters of the acquiring company. The site resisted corporate process safety initiatives, while headquarters was unreceptive to the site's needs and ideas. The report stated:

> *"The Panel believes that a good safety culture requires a positive, trusting, and open environment with effective lines of communication between management and the workforce, including employee representatives. The single most important factor in creating a good process safety culture is trust (emphasis added)."*
>
> Culture problems went well beyond *Mutual Trust*. Starting up the column without first considering the presence of nearby temporary trailers demonstrated a failure to *Understand and Act on Hazards and Risks*. Not following the written start-up procedure clearly demonstrates *Normalization of Deviance*. That the written procedure simply did not work and was never fixed suggests an absence of *Open and Frank Communication*. Finally, the incorrect use of occupational safety metrics to indicate good process safety performance is a typical symptom of no *Sense of Vulnerability*.

The American Heritage Dictionary (Ref 2.11) defines trust as:

> *"Firm reliance on the integrity, ability, or character of a person or thing; confident; faith. The person or thing in which confidence is placed."*

Successful, productive interaction between people relies on trust. Each party in the interaction needs to trust that the other means what they say and will do what they say they will do. In a trusting relationship, people believe that their superiors, peers, and reports will make the right decisions, act honestly, follow-through, and if necessary resolve conflict.

Trust is therefore the glue of process safety culture. Many of the process safety culture core principles cannot succeed without trust across the organization. Trust clearly underpins the core principle of open and frank communication, for without trust, people will not speak frankly. Trust is inherent in empowering others to fulfill their process safety responsibilities. Without trust, a leader assigning a critical process safety task accepts that it may not be done. And without mutual trust between managers and their experts, deference to expertise simply cannot happen.

In the absence of trust, employees may dismiss a manager's statements about safety as not serious. Managers may seek blame for errors, rather than seeking root causes. Workers may be reluctant to report near-misses and incipient safety problems. Managers may withhold funding for safety because they believe workers are being lazy. Managers may second-guess the experts and end up making poor decisions.

When trust is lost, a culture can be seriously damaged. Trust can be destroyed much faster than it can be built. And, rebuilding trust after it is lost can take as long as building it in the first place.

CCPS cites lack of trust as a key warning sign of potential catastrophic incidents (Ref 2.12). Table 2.1 summarizes some indicators of trust and mistrust within an organization.

Table 2.1 Indicators of Trust or Mistrust

Trust Indicators	Mistrust Indicators
Personnel willingly volunteer	"It's not my job" (Ref 2.13)
Leadership and peers trusted to make right decisions	Problems and concerns hidden from management
Problems reported without fear of reprisal	People feel they are missing essential information
Employees satisfied with problem resolution	Unusual friction between groups, shifts, facilities, corporate, etc.
Auditing and checking welcomed	Cliques
Comradery across organization	Recklessness tolerated (Ref 2.13)
Bad ideas challenged when proposed	Conflict with contractors, neighbors, labor, press, etc.

(After Ref 2.12 except as noted)

In summary, the process of developing mutual trust often starts with words – a declaration of intent or a request for cooperation. However, trust is truly created by deeds – living up to the words. Trust is a valuable but fragile commodity. It is hard to create and easy to violate, and once it has been violated it is difficult to regain. Therefore, process safety leaders should pay attention to trust and work to earn it every day.

2.4 ENSURE OPEN AND FRANK COMMUNICATIONS

Over Texas and Louisiana, USA, February 1, 2003

The space shuttle Columbia broke up upon re-entry, killing all seven crew members (Ref 2.15). During the initial minutes of flight, insulating foam detached from the external fuel tank, striking and damaging the shuttle's heat resistant tiles. Without the tiles' protection, the heat of re-entry melted the structural support of the wing. The resulting damage destabilized Columbia, and the resulting stress led to its disintegration.

The shuttle's design specification stated, "No debris shall emanate from the critical zone of the External Tank on the launch pad or during ascent." However, investigation revealed that foam loss had occurred during more than half of previous missions, in many cases damaging tiles.

Employees and contractors at several NASA sites had noted this as a concern among their local groups, but the culture discouraged *Open and Frank Communication* of this concern to management. One memorandum, composed but never sent, said, "I must emphasize (again) that severe enough damage... could present potentially grave hazards...Remember the NASA safety posters everywhere around stating, 'If it's not safe, say so'? Yes, it's that serious."

> When the large foam strike was observed on video after the launch, attempts to communicate concern were ignored. Also, suggested additional surveillance were flatly denied.
>
> Many other culture failures existed in the NASA organization leading up to the loss of the Columbia. Data on progressively larger foam strikes that did not lead to thermal damage clearly illustrate the evolution of *Normalization of Deviance*. These data were then extrapolated incorrectly, a failure to *Perform Valid and Timely Hazard/Risk Analysis*. The NASA organization had become so fixated on schedule that they any bad news, including safety, was discouraged, destroying the *Imperative for Process Safety*. Reference 2.1 discusses the culture findings in more detail.

The playwright George Bernard Shaw famously stated: "The single biggest problem in communications is the illusion that it has taken place." Leaders may believe they have delivered clear, well-crafted messages, and then be surprised later to learn that the recipients completely missed them.

Some forms of communication are required to be written and formal by regulations or company standards to serve as legal records of compliance. This section does not address such requirements but instead considers the impact of communication on culture.

Communication requires more than one person simply transmitting information towards another. The receiver of the information must also be tuned-in to the person communicating. The receiver must find the communicator believable and be convinced that the information is true, important, and reconcilable with their beliefs and values. Lastly, the communication should be confirmed so that the sender knows that the communication has been received. Non-verbal cues such as body language, vocal inflections, and facial expressions can either reinforce or contradict the message (Ref 2.12) and help confirm that the message was received as intended. This process is summarized in figure 2.2.

Figure 2.2 Anatomy of a Communication

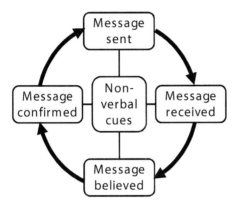

The link to trust should be clear. Trust helps the receiver of the message tune-in to the deliverer and provides the environment that encourages the receiver to provide feedback. However, the non-verbal cues of the deliverer may be even more important. If a leader does not in their heart believe in process safety, their non-verbal cues will contradict anything positive they say. Communications should of course lay out a logical argument, be unambiguous, and use a speaking or writing style that is easily understandable to the audience.

Communication should be accurate. Accurate communications avoid confusion, wasted time, and incorrect decision making, and are most easily understood. Ensuring that all communications are accurate is almost a hopeless task, given normal human error. Occasionally, people will communicate information they believe to be correct, but later turns out not to be. If a culture of open communication has been established, the incorrect information can be corrected through open and frank discussion. However, leaders must not communicate intentional misinformation, as this will eventually erode trust and make it much harder to communicate in the future.

Most people communicate good news more successfully than bad news. People naturally resist bad news. When a person receives bad news, they may deny the facts, direct anger against the messenger, or try to negotiate. Sometimes, they will do all three in succession. This behavior occurs whether messages are being delivered from a position of authority to workers, from workers to authority, or among peers.

To facilitate communication of both bad and good news, organizations should establish and maintain healthy communication channels. These should exist vertically in both directions, with managers listening as well as speaking. Vertical communication should include facility-to-corporate and facility/corporate to the board of directors.

Horizontal communications, both within and between different organizational units (hereafter referred to as silos, and including contractors, are essential for operations, and should not have to rely on chain-of-command. This helps the facility execute the elements of the PSMS and address problems. Redundant or nontraditional communications channels should be established where necessary to provide adequate communications.

In the example leading this section, NASA's organizational culture discouraged bad news. Instead, it forced "consensus" and suppressed dissenting views. Their chain-of-command approach stifled communication and diluted messages. Participants were afraid to speak up, whether to communicate the potential impact of near-misses or report actual problems (Ref 2.16).

Communication channels also need to be established outside of the reporting structure and operating groups of the facility. The facility and its parent company will normally communicate with outside organizations in a variety of situations that span from routine to crisis. These outsiders include labor, emergency responders, law enforcement, media, regulators, interest groups, and community groups, among others.

Open communication channels with outside groups must be available for use when needed, sometimes on an emergency basis. By establishing personal relationships between the parties, the foundation of trust can be established. With that relationship, it can be easier to communicate clearly in their language. In emergency situations, being able to provide the critical information needed to responders and coordinate efforts is essential. Communicating with the media during a crisis is particularly important and very sensitive (see section 4.4).

The closure of once-open communications channels can place individuals in an ethical dilemma when they believe that an important process safety message is being suppressed or failing to reach decision makers. This was a contributor to the Challenger incident (Ref 2.17) and raises ethical questions (see section 4.3).

Workplace communications can be written or verbal, official or unofficial, and formal or informal. Each has advantages and disadvantages, but all should be addressed in culture improvement efforts.

Written communication includes the many forms of electronic communication. Well-written communications, edited to improve clarity, can be less subject to misinterpretation. Verbal communication has the potential advantage of being able to fully convey feeling. If that feeling shows the speakers positive emotional commitment to process safety, the communication can be strongly enforced.

Of course, pitfalls exist with both written and verbal communication. Communications written poorly can be easily misinterpreted or simply ignored. Verbal communication can be less precise or contain errors and therefore be misinterpreted. If the emotion conveyed by a speaker differs from the message, the contradiction can cause the user to lose trust in the speaker.

Both written and verbal communication can be enhanced using pictures and drawings. Photos of past incidents can help communicate the importance of process safety. Drawings showing the correct equipment configuration can help communicate the correct way to set up the process or perform maintenance. Drawings and pictures are particularly needed when the communication is critical and/or important.

Official communications generally need to be archived, and are therefore usually in writing. That does not rule out official communications that are verbal, such as an off-hour verbal approval of a change by phone. Such communications should be documented in writing when the approver returns. Official communications typically relate to the performance of one's job. Examples include policies and standards, goals, night orders, operator logs and shift-change reports, operating procedures, maintenance procedures, and emergency response plans.

Unofficial communications are nonetheless vital ways that important information is transferred between parties. For example, a Maintenance Manager and a Lead Inspector may meet in passing and discuss how an upcoming piping inspection should be handled. Workers may also communicate informally by passing on suggestions to improve job technique or safety. In some cases, informal communication may help test ideas to eventually incorporate into a formal communication.

Unofficial communication can also be a way that positive culture spreads – or gets degraded. Consider one situation where a leader and supervisor are overheard talking about their commitment to process safety and how they plan to improve it. Compare this to another situation where the leader expresses frustration how safety rules get in the way. In both cases, the overheard conversation will be quickly and unofficially spread throughout the organization.

Formal communications tend to be official and written, while informal communication tend to be unofficial, either verbal or written. As mentioned above, official verbal communication should normally be followed up in writing to formalize it.

Informal methods can be unscheduled and ad hoc, for example, a special briefing held before an operational task or maintenance job. However, they can also be planned, such as a leader making an unannounced process safety walk-around, or attending safety meetings or toolbox talks (Ref 2.18).

Note that informal does not infer casual or of lesser importance. If the parties involved regard the channels of communication as open and the parties believe that the information is truthful, then the level of formality is not as important.

The checklist in Table 2.2 can be used to help evaluate the culture of communication.

Table 2.2 Communication Culture Checklist

Communication indicator	Observed?
Positive indicators	
Does mutual trust exist between communicating parties?	
Does a reporting culture exist that allows honest mistakes, near misses, and improvement ideas to flow freely?	
Are communications factual and never intentionally misleading?	
Are issues being communicated in a completely honest manner?	

Table 2.2 (Continued)

Communication indicator	Observed?
Positive indicators (Continued)	
Is the emotional commitment to process safety communicated broadly?	
Are images used to reinforce communications?	
Are communications confirmed to ensure they were delivered successfully?	
Negative indicators	
Are members of the organization afraid that they will face retribution if they challenge bad ideas?	
Does an us-versus-them mentality exist between shifts, departments, or production areas, management vs labor, employees vs. contractors, etc.?	
Do leaders within groups or shifts dislike each other, avoid each other, etc.?	
Do some groups or stakeholders feel like they receive insufficient communications?	

2.5 MAINTAIN A SENSE OF VULNERABILITY

Chernobyl, Ukraine, (Former) USSR, April 26, 1986

Thirty-one people died when a nuclear reactor melted down. More than two hundred suffered radiation poisoning and radioactive contamination spread over the western Soviet Union, Eastern Europe, and Scandinavia. The entire local community had to be evacuated until the damaged reactor could be encased in concrete.

During an unauthorized trial, the cooling water level decreased to the point that the recirculation pumps would not operate. As a result, the core overheated and began to

> decompose the remaining cooling water into hydrogen and oxygen. This mixture subsequently exploded, causing a loss of nuclear containment.
>
> Engineers at the facility conducted the trial hoping to find a way to more safely shut down and restart. The trial required operating with the cooling water level below the safe operating level. The engineers with a lost *Sense of Vulnerability* never considered that their trial could place the reactor in an unsafe condition. Instead they accepted a *Normalization of Deviance* and did not consider the need to *Understand/Act upon Hazards/Risks*.

Catastrophic incidents involving hazardous materials or activities happen all too often. However, they happen in relatively small numbers compared to the large number of hazardous facilities. For this reason, many facilities and indeed many companies have not experienced very large incidents. This can create a false sense of security and complacency.

The opposite of complacency – a sense of vulnerability – naturally follows a serious incident or near-miss. This can readily be understood by considering a close call while driving a car. Immediately following the incident, the driver becomes more intensely focused on the surrounding environment, looking for the next threat. They will likely drive slower, provide greater spacing from the car in front, and change lanes less frequently. This will continue for a time.

Similarly, following the Texas City explosion, many leaders of refineries and similar facilities felt a keen sense of vulnerability. They undertook many activities to improve process safety culture, management systems, and engineering.

However, people tend to forget the lessons learned from catastrophic incidents, lesser incidents, and near misses relatively quickly. A study by Throness (Ref 2.19) showed that unless people were reminded of the incident, the incident would disappear from their everyday thoughts in as little as 3 years.

Company risk management philosophies can compound the lost sense of vulnerability to process safety incidents. An overall portfolio of similar risks consists of some risks with high frequency but relatively low consequence, as well as some – like process safety – having low frequency but high consequences. Human nature tends to perceive less risk with lower frequency, even if the actual risk is identical.

In a thought-provoking series of papers, Murphy et al. (Ref 2.20) apply a convention first described by Aristotle, describing process safety incidents as "Black swans." In the northern hemisphere, essentially all swans are white, so a black swan was considered something so rare as to be virtually impossible. As Aristotle, and Murphy, point out, black swans do exist. When black swan events occur, and are then examined closely, they are found to not only be possible but probably should have always been considered white swans.

Considering potential incidents as white swans is a clever way to start developing a sense of vulnerability. However, leaders should take notice of how these swans darken to gray and eventually black as sense of vulnerability is lost, from reasons such as:

- Failing to remind personnel what happened in previous incidents on the site and similar sites,
- Incomplete investigations of incidents and near misses that do not reveal the true root causes of the incidents in question,
- Failing to share lessons learned with other company sites and the broader industrial community so that these lessons learned can be incorporated into standards and guidelines; and

- Becoming convinced that the conditions that led to the previous incident can no longer exist.

This undesirable trend has been seen many times across the industry, and even can happen repeatedly. Murphy et al. points out how overall sense of vulnerability related to facility siting has been lost many times over the years:

- Siting issues amplified the damage and casualties from the 1974 Flixborough, UK explosion (see section 2.6). Lessons learned about siting from that incident provided a strong reminder about siting to the industry.
- Siting issues amplified the damage and casualties from the 1989 Pasadena, TX, USA explosion. Lessons learned led to a new standard (Ref 2.21) and a new best practice (Ref 2.22).
- Siting issues amplified the damage and casualties from the Texas City, TX, USA explosion (Ref 2.10). Lessons learned led to one updated standard (Ref 2.23) and two new standards (Refs 2.24 and 2.25).

Will siting lead to another major incident? The above history suggests it will, and that other types of incidents will also recur, unless leaders actively promote a sense of vulnerability. Leaders also need to look for evidence that the sense of vulnerability has begun to weaken in their management, their peers, and their reports, as well as in their own minds.

In a process safety program, complacency can also develop when there is a misunderstanding of what a good process safety program means and how it is measured. For example, multiple investigations conducted by the CSB have found companies misled into complacency about process safety based on good occupational safety metrics. While it might seem reasonable to expect occupational safety performance and process safety performance to correlate, research has shown otherwise. In a study of publicly reported data, Elliot (Ref 2.26) found no statistical relationship between occupational injury and illness rate and the process safety incident rate.

Another source of complacency is the assumption that compliance with applicable regulations and standards is sufficient to abate a facility's risk. The Canadian National Energy Board (Ref 2.14), found a widespread belief that regulations must be adequate, or else they would have been made stronger (see also Appendix B). Even though many companies recognize that this belief essentially required an incident to occur before regulations could be strengthened, this reason was still used to justify a regulations-only approach. However, a company with a strong sense of vulnerability should be sensitive to all relevant risks, whether regulated or not.

The concept of High Reliability Organizations (HROs) has been used in some high-hazard fields to help improve major hazard accident safety. HROs produce products and results relatively error-free over an extended period (Ref 2.8) despite operating in hazardous conditions where the consequences of errors could be catastrophic (Ref 2.4). Two key attributes of HROs that help establish a sense of vulnerability are:

- They have a chronic sense of unease, i.e., they have a strong sense of vulnerability and avoid complacency. They know that the absence of incidents over time does not mean the next incident cannot happen imminently.
- They have strong responses to weak signals, i.e. they set their threshold for intervening very low. If something does not seem right, they stop operations and investigate. This means they accept a much higher level of 'false alarms'. Most importantly, HROs do not become complacent of false alarms and consistently evaluate why each occurred.

High-risk industries in which some facilities follow the HRO model include commercial and naval nuclear power, aircraft carrier flight operations, airlines, hospitals, and other high-risk industries. Lodal (Ref 2.27) points out that the hazards inherent in chemical, oil and gas, and related facilities have the same kind of immediacy and severity as other industries that follow the HRO model. Lodal also suggest that becoming high reliability has a positive connotation that can strengthen the sense of vulnerability.

A strong can-do attitude among facility personnel can also weaken the sense of vulnerability. Most leaders would consider this to be a positive cultural trait and an indicator of a strong esprit de corps, or personnel bonding within the organization. They may also point to their team's successes and their ability to recover from upsets, preventing severe consequences. However, this trait can result in the willingness to take chances that are not consistent with the risks at hand and organization's risk tolerance.

Investigators found that the can-do attitude of NASA contributed to both the Challenger and Columbia incidents. In the earlier Mercury, Gemini, and Apollo programs, engineers and managers observed backup systems taking over when primary systems failed; they figured out how to return Apollo 13 safely to earth despite losing many of the back-ups; and most of all, they succeeded in sending men to the moon. Such an attitude contributed to minimizing the O-ring and foam strike near-misses as problems they could work around, despite violations of safety of flight standards (Refs 2.15, 2.17).

In its deliberations regarding the process safety culture at Texas City and four related US refineries, the Baker Panel (Ref 2.5) noted that a strong can-do attitude can also result in overconfidence that encourages bypassing established procedures and practices. A sense that corporate process safety initiatives are at cross purposes with facility process safety initiatives could impair process safety can result in bypassing or ignoring of those corporate procedures. This can further result in ignoring of any outside advice and a form of self-isolation with respect to new or different process safety ideas.

Maintaining a sense of vulnerability also requires that organizations be vigilant for new or previously unrecognized causes of process safety incidents. When new issues are discovered organizations should then extend the scope and application of their PSMS to cover these new issues. Examples of such extensions are:

- Many organizations have extended the use of their Management of Change (MOC) program to include certain types of organization and personnel changes. Organizational Management of Change (OMOC) was not originally part of the intent of MOC, but many facilities and companies have recognized how turnover in certain jobs, overall staffing, and other similar changes can affect the quality of the PSMS.
- Many organizations voluntarily perform Layer of Protection Analysis (LOPA) as part of their Hazard Identification and Risk Analysis (HIRA)/Process Hazard Analysis (PHA) to provide additional study of the number and quality of their safeguards for possible hazard scenarios that meet certain risk criteria as measured in their HIRAs/PHAs.

Complacency and an uncontrolled can-do attitude are part of human nature. They can be reinforced by the social conditions within an organization, but mostly they represent human traits that are common to all people to some degree. Combatting these characteristics can be difficult, even when the risks are high. When competing pressures, such as production are also present, a can-do attitude can become a complicating negative trait.

2.6 UNDERSTAND AND ACT UPON HAZARDS/RISKS

Flixborough, North Lincolnshire, UK, June 1, 1974

A vapor cloud explosion following the failure of temporary bypass piping killed twenty-eight workers. Many other workers suffered injuries and significant onsite and offsite property damage occurred. The temporary piping had been installed to bypass the fifth oxidation reactor in a chain of six. Reactor five had failed and was being repaired.

Supported only by conventional scaffolding, the temporary piping was installed without first *Understanding and Acting Upon the Hazards and Risks*. Considering the haste to install the bypass and the close spacing of work areas on the site, the facility appeared to have a weak *Sense of Vulnerability*. After a two-month exposure to stress, vibration, and fatigue, the piping failed, creating a large release of flammable vapors.

The Flixborough incident hastened passage of the UK Health and Safety at Work Act. While it predated the development of formal PSMS elements as we know them, it remains a classic example of failures of the Management of Change (MOC) and HIRA/PHA elements. Both elements rely heavily on dedication to understanding hazards and risks, and how they can change as the process changes. Understanding hazards is also a key aspect of the PSMS element "Competency" (see section 5.4).

Leaders should understand the difference between hazards, risks, and the safeguards that are used to act on these hazards and risks. The hazard of a material is the harm it can inflict. Process hazards include toxicity, flammability, reactivity, high and low pressure, and high and low temperature. Physical impact, electrical shock, and suffocation may also be process hazards.

Hazards have potential consequences if not managed properly, such as fires, explosion, toxic effects, and environmental damage.

Risk is a function of the probability of an incident and the potential consequences if the incident occurs. Process safety incidents typically have low probability but high consequence. By contrast, occupational safety incidents typically have a relatively higher probability, but the consequences are smaller. Because consequences of process safety incidents can potentially be quite great, process safety risks could be greater than occupational safety risks.

Safeguards are applied to reduce risk to or below a standard level that the company, after deliberation, has deemed tolerable. Safeguards may reduce the potential consequences of an incident (Inherently Safer Design), probability of an incident, or both. As much as possible, safeguards must be independent such that a common failure can eliminate more than one safeguard. Generally, the HIRA process should identify engineering and process solutions to keep safeguards independent.

However, time and again, incidents are caused by one specific common failure – the failure to act upon known hazards and risks. Failures to act include not following up on action items, not keeping safeguards in place, not performing the required preventative maintenance, testing, and inspection, and not analyzing changes to the process.

In a strong process safety culture, leaders and personnel at all levels should strongly believe in understanding their process risks. They should conduct hazard and risk reviews as thoroughly as necessary to develop this understanding. Then they should implement the engineering and operational measures needed to control these hazards and manage the risk. They should carefully consider the effects of changes at the facility, which may require changes to the way these hazards and risks are managed. Finally, they should take the required measures to assure that their safeguards function as required.

In a weaker culture, PHAs and MOCs may be ignored or treated as check-the-box activities; required but not taken seriously. Action items from PHAs and MOCs may also be forgotten, ignored, or simply dismissed. But this is the road toward a serious PS incident. Without a sufficient understanding of risks and the options available to mitigate them, management cannot make effective risk decisions.

In both PHA and MOC, one size does not have to fit all. Processes having greater hazards and risks certainly deserve greater depth of study. A less formal approach may suffice for processes with lower hazards and risks. In either case, maintaining a vigilant process safety culture requires thoroughly understand the facility's process risks and then reducing those risks to an appropriate level. This tolerable level should be predetermined by the company and applied consistently (Ref 2.28).

Typical approaches for understanding process safety risks are:

- What-if and what-if checklist analysis,
- Failure modes and effects analysis (FMEA),
- Hazard and operability analysis (HAZOP),
- Layer of Protection Analysis (LOPA).
- Quantitative risk analysis (QRA); and
- Bowtie analysis (combination of QRA and FMEA).

Such analyses should be conducted throughout the lifecycle of the process, and whenever changes to the process are considered. CCPS addresses the choice of appropriate hazard analysis technique in Guidelines for Hazard Evaluation Procedures (ref 2.29).

In strong process safety cultures, leaders should question the estimates of incident probability and consequences, as well as the risk reduction enabled by safeguards made by hazard and risk analysis teams. This questioning should strive only to build understanding. It should not be seen as a second-guessing exercise. If an incident has not occurred in some time, people come to think of such events as impossible or practically so. This effect can be exacerbated in organizations that have already made considerable efforts to control process hazards, through a desire to declare their efforts complete. Consequences can become underestimated similarly. This can happen especially when safeguards have been successful in turning a potentially serious event into a minor event or near-miss.

Many companies address this issue in their hazard/risk assessment/management standard using a risk matrix. Along one axis of the matrix, they clearly define roughly order-of-magnitude levels of probability (or frequency). For example, a probability of 0.1 might be assigned to a probability of daily to weekly, and a probability of 0.0001 might be assigned to something that could happen once in the life of the process.

Similarly, along the other axis of the matrix, consequences are divided into roughly order of magnitude categories. Consequence definitions ranging, for example, from mild to catastrophic are defined. Usually, it is necessary to define consequence levels in several categories such as injuries/fatalities, property damage, offsite consequences, business impact, and environmental impact.

The matrix itself is then divided into categories of risk. Each cell represents a level of risk that is the probability times consequence of that cell. Companies will generally define a level of risk that is too high to be accepted, intermediate levels of risks that need to be addressed at a priority dictated by risk, and a level of risk that is generally acceptable. Figure 2.3 shows an example matrix.

Figure 2.3 Example risk matrix

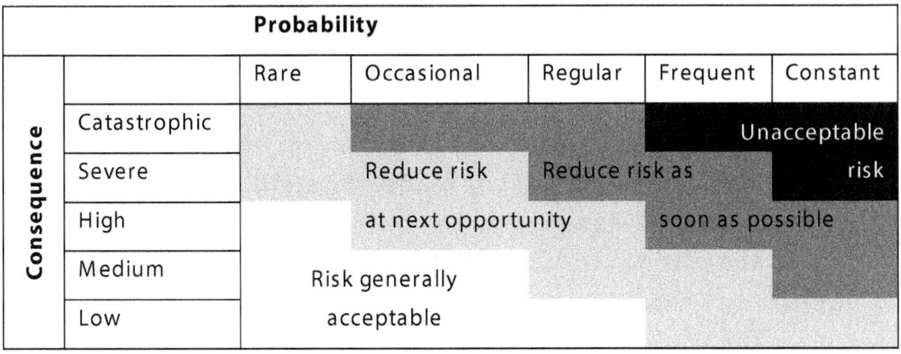

Probability						
		Rare	Occasional	Regular	Frequent	Constant
Consequence	Catastrophic					Unacceptable
	Severe		Reduce risk	Reduce risk as		risk
	High		at next opportunity		soon as possible	
	Medium		Risk generally			
	Low		acceptable			

If the risk related to a given hazard is not within the generally acceptable category, the company must then apply safeguards to reduce the risk. Again, efficacies of given safeguards are clearly defined in order of magnitude categories. Safeguards that reduce probability by 1 order of magnitude shift the risk one cell to the left in the matrix. Safeguards that reduce potential consequences by 1 order of magnitude downwards. It may take several safeguards to bring the risk to the acceptable level.

Among other benefits, the risk matrix approach makes it quite clear how many safeguards are required. Generating risk matrices can be hard work, however. It helps in defining risk categories to relate risk levels for the process to risk levels in daily life, such as the risk of driving, to help everyone can clearly see how the process risk compares to something they are familiar with.

The bottom line of understanding and acting on hazards and risks, as Admiral Hyman G. Rickover stated on many occasions, to "face the facts." As Adm. Rickover built the US Navy's nuclear program, he strongly believed that officers managing the program must be prepared to make difficult decisions that favor reactor safety, despite pressures due to cost, manpower, schedule, or potential bad press involved. Ultimately, the facts about process safety come from hazard and risk assessment. Facing these facts requires acting on them in a way that justifiably manages the risk.

Evidence of gaps in understanding and acting on hazards and risks can often be seen in partial or incorrect solutions. In one case, a facility included in its emergency plan that the evacuation muster point would be in an indoor location. They based their decision on convenience of the location and ease of quickly performing the headcount. While valid objectives, by choosing a site where indoor air could not be isolated from any potential contaminants outdoors, they potentially exposed a greater number of workers and actually increase risk.

A second example of gaps can be found in the Illiopolis incident discussed in section 2.1. Although not a contributor to the incident, the intended use of the emergency air hose was to allow operators to mitigate a runaway reaction by draining the reactor contents on the floor. If that emergency procedure ever had to be used, the consequences would have been quite similar to what happened in this incident.

2.7 EMPOWER INDIVIDUALS TO SUCCESSFULLY FULFILL THEIR PROCESS SAFETY RESPONSIBILITIES

North Sea, Offshore of Aberdeen, Scotland, UK, July 6, 1988

The explosion of the Piper Alpha platform resulted in 167 fatalities. A major release of liquified gas occurred from a metal plate temporarily installed in place of a relief valve removed for service (Ref 2.30). The incident escalated significantly due to backflow from nearby platforms that could have been shut off. However, platform managers who were not *Empowered to Fulfill Their Safety Responsibilities* did not do so because they lacked the authority and had difficulty communicating with their management on shore.

The incident occurred after several failures of *Open Communications* related to the Permit to Work and

Lockout/Tagout processes allowed operators to restart a hydrocarbon pump that had been taken out of service for maintenance on a connected relief valve. A weak *Imperative for Process Safety* allowed firewater pumps to be left off, even though there were no divers in the water, and allowed workers to be on the rig without up-to-date evacuation training.

In the 1920s, researchers at Western Electric Co. discovered that group psychology is a greater factor in employee job performance than individual psychology. Researchers also noted that the motivation of the employees determined their output more so than management edict or even fatigue. Organizations are formal arrangements of functions and responsibilities, but they are also social systems (Ref 2.31). As a result, the motivation of facility groups and sub-groups is just as important as individual motivation in a sound process safety culture, particularly in empowering personnel to successfully fulfill their process safety responsibilities.

In an empowered process safety culture, safety knowledge and leadership are both top-down and bottom-up. With the imperative for process safety established and with leaders fostering mutual trust, everyone in the organization can be motivated to comply with the PSMS and operational rules that they have had a role in creating.

In such an organization, leaders can and should provide clear delegation of process safety-related responsibilities. Along with the delegation should come the training required for the job and the necessary authority and resources to allow success in their assigned roles. Personnel accept and fulfill their process safety responsibilities. Finally, while leaders delegate responsibilities, they retain accountability. In this way, leaders delegate, but do not abdicate responsibility.

Employee participation appears as an element in many national process safety regulations. Voluntary PSMSs including CCPS's RBPS and the International Chemistry Council's Responsible Care® also embrace employee participation. Proactively consulting with workers in both the development and implementation of the process safety program is a good first step in empowerment.

Empowerment has particular importance during non-routine situations. These could be minor, such as equipment that needs maintenance ahead of schedule, or could be an obvious emergency. At either extreme and along the entire spectrum, accountabilities and responsibilities for process safety should be clearly established and documented. People at all levels of the organization should have, and know their accountabilities and responsibilities.

Care must be taken to avoid silo-ing of process safety responsibilities and information. Whenever there is a possibility to question whether one organization or another owns the responsibility, the empowered employees should collaborate to ensure that the information is shared, and responsibilities have been carried out. In this positive atmosphere of empowerment everyone should recognize that good ideas do not recognize organizational boundaries.

Stop-work authority is a vital component of empowerment to fulfill process safety responsibilities. Stop work authority covers a spectrum of possible situations, from routine operational or maintenance tasks when personnel detect a safety issue developing, to full authority to initiate an emergency shutdown of the process when warranted by the operating conditions. The use of stop work authority could have stopped the Macondo incident from proceeding to its actual conclusion after faulty mudding results were reported. Facility and company policies and procedures should fully respect this empowerment.

Employees should feel that they can stop any activity when they notice a potential hazard, even when stopping may have an impact on production or costs. They should feel that these actions can be taken without retribution from either fellow workers or management, and that second-guessing from any party regarding the consequences of such actions will not occur.

Empowerment promotes feelings of self-worth, belonging and value. Employees should be involved in training, should be consulted about the content of the PSMS, and should participate to the extent possible in all process safety activities.

2.8 DEFER TO EXPERTISE

Atlantic Ocean, Offshore of Florida, USA, January 28, 1986

Shortly after launch, the external fuel tank of the space shuttle Challenger exploded, dooming the shuttle and all 7 crew members. The fuel tank was breached by a sizeable leak of hot gases through two O-rings that sealed a joint in one of the orbiter's two booster rockets. On the day of launch, the temperature was significantly colder than the O-rings were designed to seal. However, NASA management failed to *Defer to the Expertise* of the booster rocket program engineer and launched anyway.

The investigation revealed a significant *Normalization of Deviance* (Ref 2.32), as NASA launched shuttles at colder and colder temperatures, accompanied with greater and greater burn-through of the O-rings.

The Challenger space shuttle disaster of January 28, 1986, was a major turning point in the consideration of culture in highly technical operations. However, NASA failed to *Learn to Assess and Advance the Culture*, leading to the loss of the shuttle Columbia from another *Normalization of Deviance* situation.

Deferring to expertise is a natural outgrowth of empowering individuals to fulfill their safety responsibilities. Certainly, in an empowered organization, the process of delegating responsibility involves assuring that the person assigned has the requisite knowledge and skill. This culture core principle addresses the wide range of subject matter experts, whether within the facility, in a corporate role, or extremal to the company, e.g. consultants, vendors, or academic experts.

The Competency element of CCPS's RBPS (Ref 2.33) management system specifies that an organization should maintain a sufficient level of expertise required for safe operations. Many aspects of a successful process safety program rely on specialized subject matter expertise. The employer should validate that the individual has the required education, and training and has proven their skills through experience. Once done, the leaders should then defer to those experts when technical questions arise regarding process safety.

Technical expertise can and indeed should go beyond engineering and other chemical and physical sciences. Functions such as operations, maintenance and inspection, and emergency response, among others, all have special and different expertise that deserves consideration when technical questions arise regarding process safety.

Technical expertise should include knowing what you do not know. The organization should know where its knowledge is adequate and where it is not. When a facility needs expertise that it does not have, it should seek expertise from the corporate level, and if necessary go outside the organization to find an outside expert. Critical process safety decisions should not be made with incomplete technical information. The facility can either wait until answers can be found, or apply sufficient alternative independent safeguards.

For example, many PSMSs provide a means for approving the bypass of safeguards. These variances are typically approved at a senior level, such as facility manager or operations manager. If the person holding this title is not technically competent to properly evaluate the request, they should enlist someone with the required expertise before deciding whether to approve or decline. It can sometimes feel threatening to a leader to admit that they do not have the necessary expertise. However, by demonstrating the deferral to expertise, the leader helps establish that this behavior is safe within the organization.

Deferral to expertise does not mean that facility and operations leaders can be ignorant of process safety. Indeed, such leaders should know the PSMS well and should know the key hazards and safeguards of the process. A strong process safety culture such as characterized by High Reliability Organizations (Appendix D) places a high value on the technical competence of managers who have demonstrated the prerequisite knowledge and qualifications. This basic technical competence helps leaders know what they do not know.

Process safety experts, whether in the line organization or a separate group, should maintain their competency through training and experience. Many specialized sub-skills within process safety require certifications, which should also be maintained. Process Safety experts should be accorded equal status, authority, and salary as other operational roles (Ref 2.14).

2.9 COMBAT THE NORMALIZATION OF DEVIANCE

Bhopal, Madhya Pradesh, India, December 3, 1984

Nearly 40 tons of methyl isocyanate (MIC) escaped from a pesticide plant, killing more than 3,000 people and injuring more than 100,000 in the ensuing days. In the early morning hours, water contamination of a storage led to runaway reaction causing the tank to overpressure and vent. In the weeks leading up to the incident, *Normalization of Deviance* occurred as the route for contamination was established via a jumper line. Then one-by-one, safeguards were bypassed, setting the stage for tragedy.

The parent company had decided the previous year that the facility could not operate the process safely, and, following the *Imperative for Process Safety* had decided to shut it down. However, the site leadership provided anything but *Strong Leadership* a as they shut down the facility unit by unit, failing to *Understand and Act on Hazards/Risks.*

In the absence of *Open and Frank Communications* with the community, emergency responders may have unintentionally worsened the community impact, and hospitals had neither the knowledge nor resources necessary to treat exposed persons.

Humans naturally seek to continuously improve their environment and condition. Humans invent, tweak, and optimize, always seeking something faster, cheaper, and better. Our well-being today exists largely to the inventiveness of our ancestors, and today's business climate only encourages people to continue this trend.

However, when it comes to process safety, the tendency to optimize can betray us. Process and procedure shortcuts can erode safety margins and defeat safeguards. Changes, even changes intended to improve safety (see Section 2.5), can introduce new hazards worse than those in the original process. When this happens in process safety, it is called normalization of deviance, defined as a gradual erosion of standards of performance because of increased tolerance of nonconformance.

The normalization of deviance can occur in operating, maintenance, engineering, management and other functions as well as in the condition of equipment. Over time, if nothing bad happens, facility/company personnel will naturally gravitate to thinking deviance are acceptable. The deviance simply become part of the facility operation. However, the pattern of deviance really means that the margin of safety embedded in the process/equipment design and operational rules and practices has been reduced.

Because of this, normalization of deviance may be the most common, though often unrecognized, process safety cultural deficiency. Normalization of deviance occurs subtly over time when individuals, work teams, and entire organizations gradually accept a lower or different standard of performance that de-facto becomes the norm. In this way, conditions slowly change and erode over time. The deviance does not appear to affect the process in terms of quality, yield, or productivity, except potentially in a positive manner.

To summarize, normalization of deviance requires three things:

- Initiated by a person or persons,
- Occurring repeatedly over time; and
- Not causing immediate incidents or noticeable process effects.

This does not necessarily mean a progressively large deviation in the same parameter or activity, although this is commonly associated. Normalization of deviance can also happen when various kinds of deviations occur to the same process. This might for example include skipped steps, workarounds, and violation of operating limits. Deviations do not have to be practiced consistently. For example, one work crew may have deviate from procedure in one way and a second crew in another.

The following paragraphs describe typical normalization of deviance situations. Usually, it does not take much effort to detect them. CCPS (Ref 2.12) identifies these situations as readily observable warning signs of impending incidents, if an effort is made to look for them. All the following situations have the common feature of a deviance from what had previously been deemed to be a safe means of operating and maintaining a process. As deviations become more and more normalized, the risk of a catastrophic incident increases, often well above the company's risk tolerance.

Operating outside the defined safe operating limits

Processes and equipment designs usually base process risk management on limits established through fundamental engineering, supplemented by the hazard and risk assessment process. Operating outside these limits for any reason increases the risk of the entire operation. Tolerance for routine excursions outside safe operating limits is a prime example of normalization of deviance. Workers should understand the importance of controlling the operation within allowable limits and be accountable for not exceeding these limits.

Safety systems/features out-of-service beyond time limits

Safety systems/features that remain out-of-service_beyond the time limits specified represents a normalization of deviance, even if these extensions were in accordance with policy. Such safety system removals can become quasi-permanent, although the risk may be unacceptable.

Nuisance alarms, ultimately ignored

Alarms represent critical deviations that operators should address promptly. As such, the number of alarms should be managed, either by removing less important alarms or addressing underlying process problems causing the alarms. When alarms come to be considered nuisances alarms, operators may start to ignore all alarms, even the ones signaling critical deviations.

Unreliable instrument readings, ultimately ignored

As the unreliable instrument readings increase due to improper maintenance and calibration, operators tend to ignore the readings or hesitate to shut down or report the problem. Instead they need to consider what if the reading is actually correct and address that potential risk.

Chronically overdue maintenance, testing, and inspection

The facility's risk is also based on the presence of safeguards and the integrity of the equipment. Safeguard and equipment integrity depends on performing inspection, testing, and preventive maintenance (ITPM) tasks as scheduled. Normalization of deviance occurs when these tasks become chronically overdue and/or the list of overdue tasks or length of time overdue is growing.

Growing list of equipment deficiencies

Normalization of deviance is also occurring when the list of equipment deficiencies continues growing and/or when the length of time deficient increases. This situation could drive additional normalization of deviance as workers seek patches and workarounds.

Growing list of action items

A continually growing list of action items and/or an increasing length of time overdue represents another normalization of deviance scenario. As the number of action items swell, the facility tends to become overwhelmed and give up.

Failure to follow procedures

Another common normalization of deviance scenario occurs when workers do not follow approved operating, maintenance, or other procedures, particularly when supervision is not present. Failure to follow procedures includes taking short-cuts and doing the work differently than specified. In this scenario, there may be peer pressure for coworkers and new hires to take such shortcuts. Failure of engineers and managers to follow required standards is another form of failure to follow procedures.

Different practices between shifts and teams

Some companies will allow shift supervisors and team leaders to have some operational latitude. This can be acceptable, providing it is kept within the bounds of practice accepted by the company. However, when practices deviate outside of accepted bounds, deviance can be normalized. Critical operations and practices must be maintained within the acceptable limits or systems.

Potential conflicts of interest

Typically, PSMSs are designed to avoid conflicts of interest. Initiators of a change request would generally not conduct the MOC analysis or approve the MOC. For example, operation management should generally not approve its own request to defer maintenance, and design engineers for a process should generally not lead their own PHAs.

Operator rounds become "check-the-box" exercises

Many processes require operators to make periodic rounds to check for equipment status, record certain data, and notice equipment condition, smells, drips, noise, etc. When operators check the boxes without having actually done the rounds, or record abnormal conditions on the log sheet without taking corrective action or bringing the problem to the attention of supervision. this normalization of deviance can allow problems to exist and escalate undetected.

Poor shift change meetings and maintenance hand-overs

Over time, shift changes can become incomplete, informal, or completely skipped. The Piper Alpha incident is a prime example of where an opportunity to communicate an incomplete task cascaded into a catastrophe.

Poor or declining housekeeping, increasing steam/water leaks

Often when housekeeping are deficient or declining, other aspects of plant operation are deficient or declining also. Poor housekeeping can also lead directly to incidents, for example through accumulation of combustible material, loss of labeling, degraded insulation, etc. It is hard to establish an imperative for safety when management does not otherwise take pride in the facility. Non-hazardous leaks such as water, air or steam are culture indicators similar to housekeeping. While they may be perceived as a low risk loss of containment, they indicate a deeper cultural deficiency that can lead to more serious loss of containment events.

Normalization of deviance from PSMS element requirements

Like deviance from procedures and standards (see above), deviance from essentially every PSMS element can be normalized, sometimes in multiple ways. For example:

- Deviating from the standard definitions of probability, consequence, and independent protection layers in HIRA/PHA to make risk appear lower than actual. Also, oversimplifying the process and failing to look at the entire process.
- Stretching the timeline of required periodic tasks such as ITPM, audits, and PHA revalidations, including stretching a

December task to January and claiming it was performed in both years.

- Deviating from definition of replacement-in-kind to avoid MOC, or understating the risk to obtain approval at a lower level.
- Not consistently investigating near misses and incidents thoroughly and formally, hiding incidents, or blaming the operator instead of finding root causes.
- Deviating from the safe work permit process. Doing haphazard evaluations, leaving required fields blank, not obtaining signatures, and failing to close out permits.
- Required training activities that are chronically overdue. Like any other overdue process safety related activity, overdue training indicates lack of attention and priority to required activities.
- Continuing to rely too heavily on computer based training (CBT) when other forms of training are needed to transfer knowledge reliably.
- MOC delays that either hold up the evaluation until the information is no longer valid, or cause the facility to implement the change without a proper MOC.
- Open and overdue items from MOC or PSSR.

If humans are inclined to deviate from standard practice to seek improvement of their condition, and if there are many ways to deviate from the actions designed to operate process safely, what should leaders do?

The key to combatting normalization of deviance is to be highly alert to deviations, and then provide timely responses to these deviations. Time is important because there is often only a brief period between the recognition of deviance and its escalation into an incident. The timely response to process safety issues and concerns is also part of the core principles *Understand*

and Act Upon Hazards/Risks and *Learn to Assess and Advance the Culture.*

To combat normalization of deviance, some companies seek to instill a keen sense of Operational Discipline (Refs 2.33, 2.34), i.e., measures to ensure the performance of all required tasks correctly every time. Operational Discipline may appear to apply only to operators and mechanics, but in fact it applies to all employees, including leadership. Leaders also have policies they must adhere to and tasks they must perform. In operational discipline, leaders have the added responsibility to assure that the people they manage are performing their required tasks.

Establishing high standards of process safety performance is an important aspect of operational discipline. These standards should be stressed to new employees in their initial training, and then reinforced during refresher training and regularly on the job. Where changing circumstances warrant, standards are carefully modified, but the normalization of deviance with respect to these standards is not accepted, and there is zero tolerance for willful violations of process safety standards, rules, or procedures. Zero tolerance should be clearly and quickly demonstrated when violations occur. Effective Operational Discipline requires employees at all levels to hold each other accountable for their parts of the PSMS.

2.10 LEARN TO ASSESS AND ADVANCE THE CULTURE

Longford, Victoria, Australia, September 25, 1998

Brittle fracture of a heat exchanger at a gas processing plant led to an explosion and fire that killed two people and injured several more (Ref 2.35). The damage caused the facility to shut down, leaving homes and businesses in the city of Melbourne and the state of Victoria with no natural gas supply for weeks.

Investigators found that audits conducted of the facility shortly before the incident revealed no deficiencies in the management system. These audits gave management false confidence in their process safety performance and culture, and they failed to *Learn to Assess and Advance the Culture* identified no deficiencies.

Particularly troubling was degree to which the plant failed to Understand and Act on Hazards and Risks, both in performing PHAs and MOCs. Investigators also noted indicators of other cultural deficiencies indicators, such as alarms that sounded too frequently, indicating a *Normalization of Deviance*.

Few if any companies have an ideal process safety culture. In fact, developing and maintaining process safety culture is a continuing journey. Most companies either have, or plan to make some initial improvements. These efforts should then be followed by additional culture improvements to help make the process safety culture more and more robust over time.

Moreover, the same human forces that cause deviance from procedures and standards to be normalized can also cause culture to weaken. Small deviations from commitment, leadership, trust, and so on can accumulate, ultimately undoing prior efforts to improve culture. Changes in personnel can have a similar effect, especially if competency is compromised.

Organizations that do not internalize and apply the lessons gained from mistakes, including others' mistakes, will fail to advance the culture. They are likely to relegate themselves to static, and more likely declining, levels of performance. Process safety excellence requires the curiosity and determination necessary for the organization to be a learning, advancing culture. Knowledge, communications, and a questioning/learning environment are the key characteristics of a process safety culture that constantly refreshes itself. The following paragraphs suggest some ways to learn and advance the culture.

Be adaptable

Adaptable organizations continually adopt new and improved ways to do work, and the different units or groups in these organizations often cooperate to create change. Adaptable organizations also demonstrate a strong user/customer focus. This focus on change and improvement can also be directed to improving culture. However, without careful review of proposed changes, being adaptable can lead to normalization of deviance. Therefore, culture improvement efforts should be accompanied with careful review. The organization should define and understand its boundaries of acceptable process safety performance. Any variation should keep within these boundaries (Ref 2.3).

Be competent

Each person whose job addresses process safety in some way should possess the required knowledge and skills to perform this position (Ref 2.33). Clearly, process safety experts need to be competent in their disciplines, but competence does not end there. Process safety competence applies to everyone in the facility. Design engineers should be competent in applicable standards. Decision-makers should understand how to interpret risk assessment data. Operators and mechanics should be able to perform all required tasks and understand the importance of following procedures. Leaders should understand and be able to build process safety culture.

Be aware

All facility personnel should support the process safety program, whether they have a formal process safety role or not. Each person on the site should be thoroughly indoctrinated into the precepts of the PSMS and how it applies at the facility. Each person should also know their role, even if it only to evacuate promptly to the specified location when the alarm sounds. Awareness should go beyond a statement of the sites' process safety principles, and should include incident case histories to help maintain the joint sense of vulnerability.

Communicate

Keeping vertical and horizontal communications channels open, whether discussing process safety or any other subject, encourages continuing honest and open communications. Communications help alert leadership to potential problems, promote opportunities for improvement, and provide an informal way to assess the status of the culture. When problems are communicated, the messenger should be rewarded, not punished.

Question and learn

A questioning and learning environment helps companies adapt, communicate, and improve. Facility personnel become comfortable asking "why?" and find their coworkers happy to provide answers. Likewise, constructive comments are freely offered and gratefully accepted. People do not necessarily know how to offer comments and pose questions positively, so training and development in this skill may be needed. Developing this skill does help avoid feelings of being constrained to offer opinions or fear of being embarrassed or insulted from posing a question.

Find and address root causes

The organization should recognize that the causes of catastrophic events and near misses are usually complex. Consequently, there is rarely a single simple solution to a process safety problem, and operating error is never the only cause of an incident (Ref 2.33). For example, the response to a corrosion-caused release should not stop with just replacing the corroded pipe. Was the corrosion rate excessive? If so why? Does this suggest a processing problem that must be addressed? Why was accelerated corrosion not detected? Bear in mind that most failures and near misses are rarely a unique event that can be shrugged off. Instead, fix the root causes to potentially uncover a broader more invasive problem.

Be sensitive to leading indicators

Formal leading indicators (see Section 3.1) can help identify normalization of deviance and other developing problems with the PSMS and process safety culture. In the post mortem of nearly all undesired events, the investigations revealed that the information needed to detect, prevent, or mitigate the events in question were available to the organization but they were ignored or not understood. Near-misses are potent leading indicators that should not be ignored, because they highlight conditions that are more likely to cause an incident. See Appendix D for a discussion of how high reliability organizations act on leading indicators and near-misses.

Assess the culture

Part of assessing the process safety culture of an organization involves formally measuring it periodically. Organizations should establish carefully considered and mutually agreed upon key performance indicators that should be collected, reported, and analyzed by organization management on a periodic basis. See Section 4.5 for a more detailed discussion or process safety culture metrics. See section 4.2 for a discussion of the relationship between process safety culture metrics and compensation.

Know the subcultures

While the ultimate goal is to develop a single culture that applies broadly across the company, subcultures can exist within the organization. Process safety cultures can differ between work groups and shifts in a facility, between unions and management, among others. A survey of nine hourly and salaried work groups in a refinery (Ref 2.5) clearly showed culture differences between the groups and a wide divergence in responses between workers and management. Advancing culture under such a situation may require initially addressing each of the subcultures differently before they can be moved to the common desired culture. Also, the diversity provided by subcultures can also be a source of opportunity in culture improvement efforts, both in terms of helping identify problems as well as providing a range of positive examples.

<u>Exercise patience in culture changes</u>

Changing and improving PS culture is like turning a large ship; it starts with a decision for a new course, takes a long time to reach the new heading and requires continued effort to maintain that direction. Leaders need to realize that culture changes take months if not years to become fully implemented. Our tendency is to expect prompt progress

toward a new goal. If the culture message is not consistent across the organization and across time, it will be marked as a passing fad and the opportunity for lasting cultural change will be lost.

2.11 SUMMARY

The core principles described in this chapter describe process safety culture in a high-level roadmap to culture and how to improve it. The first three principles (e.g. Establish an Imperative for Process Safety, Provide Strong Leadership, and Foster Mutual Trust) provide a necessary foundation for implementing the other seven principles.

However, improvements in all ten core principles can be gained at any time in the culture improvement effort. Notably, failures in maintaining a sense of vulnerability and combatting the normalization of deviance have proven quite common contributors to incidents. Therefore, special emphasis on these two core principles can yield low-hanging fruit.

Implementing these principles in practice, with behaviors and concrete policies and procedures will be discussed in Chapters 4, 5, and 6.

2.12 REFERENCES

2.1   CCPS, Process Safety Culture Tool Kit, American Institute of Chemical Engineers, New York, 2004.

2.2   Chemical Safety and Hazard Investigation Board, *Investigation Report—Vinyl Chloride Monomer Explosion, FPC USA Illiopolis, IL, April 23, 2004*, March 2007.

2.3   Musante, L. et al., *Doing Well by Doing Good: Sustainable Financial Performance Through Global Culture Leadership and Operational Excellence*, Echo Strategies, October 2014.

2.4   UK HSE, *High Reliability Organisations – A Review of the Literature*, Research Report HR899, 2011.

2.5   Baker, J.A. et al., *The Report of BP U.S. Refineries Independent Safety Review Panel*, January 2007.

2.6   Reason, J., *Achieving a Safe Culture: Theory and Practice*, Work & Stress, Vol. 12, No. 3, 1998.

2.7   Chemical Safety and Hazard Investigation Board, *Investigation Report – Fire and Explosion at the Macondo Well, Deepwater Horizon Rig, Vol. 2, April 20, 2010*, April 20, 2016.

2.8   Organisation for Economic Co-operation and Development (OECD), *Corporate Governance for Process Safety Guidance for Senior Leaders in High Hazard Industries*, 2012.

2.9   CCPS, *A Call to Action Next Steps for Vision 20/20*, New York, 2014.

2.10 Chemical Safety and Hazard Investigation Board, *Investigation Report—Refinery Explosion and Fire, BP Texas City, Texas, March 23, 2005*, March 20, 2007.

2.11 Morris, W., et. al., *The American Heritage Dictionary of the English Language, New College Edition*, Houghton Mifflin Co, 1976.

2.12 CCPS, *Recognizing Catastrophic Incident Warning Signs in the Process Industries*, American Institute of Chemical Engineers, New York, 2012.

2.13 Mathis, T., Galloway, S., *STEPS to4Safety Culture Excellence^{SM}*, Wiley, 2013.

2.14 Canadian National Energy Board (CNEB), *Advancing Safety in The Oil and Gas Industry - Statement on Safety Culture*, 2012.

2.15 National Aeronautics and Space Administration, *Columbia Accident Investigation Board Report*, Washington, DC, August 2003.

2.16 Jones, D., Kadri, S., *Nurturing a Strong Process Safety Culture*, Process Safety Progress, Vol. 25, No. 1, American Institute of Chemical Engineers, 2006.

2.17 Rogers, W.P. et al., *Report of the Presidential Commission on the Space Shuttle Challenger Accident, Washington, DC*, June 6, 1986.

2.18 UK HSE Health and Safety Laboratory, *Safety Culture: A Review of the Literature*, HSL/2002/25, 2002.

2.19 Throness, B, *Keeping the Memory Alive, Preventing Memory Loss That Contributes to Process Safety Events*, Proceedings of the Global Congress on Process Safety, 2013.

2.20 Murphy, J., Conner, J., *Black Swans, White Swans, and 50 Shades of Grey: Remembering the Lessons Learned from Catastrophic Process Safety Incidents*, Process Safety Progress, American Institute of Chemical Engineers, 2014.

2.21 American Petroleum Institute, *Management of Hazards Associated with Location of Process Plant Buildings*, API RP-752, 1st Ed., 1995.

2.22 CCPS, *Guidelines for Evaluating Process Plant Buildings for External Fires and Explosions*, American Institute of Chemical Engineers, 1996.

2.23 American Petroleum Institute, *Management of Hazards Associated with Location of Process Plant Permanent Buildings*, API RP-752, 2nd Ed., 2009.

2.24 American Petroleum Institute, *Management of Hazards Associated with Location of Process Plant Portable Buildings*, API RP-753, 1st Ed., 2012.

2.25 American Petroleum Institute, *Management of Hazards Associated with Location of Process Plant Tents*, API RP-756, 1st Ed., 2014.

2.26 Elliot, M., et. al., *Linking OII and RMP Data: Does Everyday Safety Prevent Catastrophic Loss?* International Journal of Risk Assessment and Management, Vol. 10, Nos. 1/2, 2008.

2.27 Lodal, P., *Taking the FUN Out of Process Safety*, Chemical Engineering Progress, American Institute of Chemical Engineers, 2014.

2.28 CCPS, *Guidelines for Developing Quantitative Safety Risk Criteria*, American Institute of Chemical Engineers, 2009.

2.29 CCPS, *Guidelines for Hazard Evaluation Procedures, 3rd Ed.*, American Institute of Chemical Engineers, 2008.

2.30 HM Stationery Office, *The Public Inquiry into the Piper Alpha Disaster*, Cullen, The Honourable Lord, 1990.

2.31 Roughton, J, Mercurio, J, *Developing an Effective Safety Culture: A Leadership Approach*, Butterworth Heineman, 2002.

2.32 Vaughn, D., *The Challenger Launch Decision: Risky Technology, Culture and Deviance at NASA*, The University of Chicago Press, 1996.

2.33 CCPS, *Guidelines for Risk Based Process Safety*, American Institute of Chemical Engineers, 2007.

2.34 CCPS, *Conduct of Operations and Operational Discipline: For Improving Process Safety in Industry*, American Institute of Chemical Engineers, 2011.

2.35 Hopkins, *Lessons from Esso's Gas Plant Explosion at Longford*, Australian National University, 2000.

3
LEADERSHIP FOR PROCESS SAFETY CULTURE WITHIN THE ORGANIZATIONAL STRUCTURE

3.1 DEFINITION OF PROCESS SAFETY LEADERSHIP

As summarized in section 2.2, establishing and maintaining a strong positive process safety culture requires leadership. Process safety leadership does not appear without effort, nor does it exist in a vacuum. The qualities displayed by strong process safety leaders mirror the qualities of leaders in general, differing only in focus. In each situation, leaders inspire by socializing their influence and earn the respect of the colleagues and subordinates.

The defining personal characteristics and technical skills required to be a manager differ from those required to be a leader. To succeed in their process safety duties, facility managers must also have the necessary process safety leadership skills.

The following sections first briefly defines leadership in general, with references to some of the vast amount of literature available. Then these general concepts of leadership are applied to process safety.

Essential Practices for Creating, Strengthening, and Sustaining Process Safety Culture, First Edition. CCPS. © 2018 AIChE. Published 2018 by John Wiley & Sons, Inc.

Leadership in General

For centuries, scholars have maintained a continuing study of leadership and leaders and have written volumes about this topic. Much of this research has focused on the psychology of leadership. Leadership has been described as "a process of social influence in which a person can enlist the aid and support of others in the accomplishment of a common task" (Ref 3.1). Some understand a leader simply as somebody whom people follow, or as somebody who guides or directs others. Others define leadership as organizing a group of people to achieve a common goal. Studies of leadership have produced theories involving traits, situational interaction, function, behavior, power, vision and values, charisma, and intelligence, among others (Refs 3.1,3.2).

The study of the characteristics or traits that distinguish individuals as leaders continues today. Early researchers pursuing this line of research theorized that leadership is rooted in the characteristics that certain individuals possess. This came to be known as the trait theory of leadership. In other words, leaders are born, not developed.

In the late 1800's Rhodes believed that public-spirited leadership could be nurtured. This meant first finding young people with "moral force of character and instincts to lead" (Ref 3.3). Then, these young people should be educated in the proper environment such as the University of Oxford to further develop such characteristics. This vision of leadership underlays the creation of the Rhodes Scholarships, which have helped to shape notions of leadership since their creation in 1903. In other words, Rhodes believed that leaders are born, but must be developed to reach their potential. Note that at this time in Great Britain, only the aristocracy attended university; they were literally born to be the future leaders.

In the late 1940s and early 1950s, however, a series of qualitative reviews (Refs 3.4, 3.5. 3.6) prompted researchers to take a drastically different view of the driving forces behind leadership. The overall evidence suggested that persons who are leaders in one situation may not necessarily be leaders in other situations. The focus then shifted away from traits of leaders to an investigation of the behaviors of leaders that were effective. This approach dominated much of the leadership theory and research for the next few decades.

During the 1980s, advances in the application of statistics allowed researchers to quantitatively analyze and summarize the findings from a wide array of studies. This allowed trait theorists to create a comprehensive and quantitative picture of previous leadership research. Equipped with new methods, leadership researchers revealed the following (Refs 3.7, 3.8, 3.9, 3.10, 3.11, 3.12):

- Individuals can and do emerge as leaders across a variety of situations and tasks.
- Significant relationships exist between leadership emergence and such individual traits as:
  o Intelligence,
  o Ability to Adjust,
  o Extraversion,
  o Conscientiousness,
  o Openness to experience; and
  o General self-efficacy

The Integrated Psychological Theory began to attract attention after the publication of Scouller's Three Levels of Leadership model in 2011 (Ref 3.13). Scouller argued that the older theories offer only limited assistance in developing a person's ability to lead effectively. He pointed out, for example, that:

- Trait theories, which tend to reinforce the idea that leaders are born not made, might help in the selection of leaders, but they are less useful for developing leaders.
- One ideal leadership style would not suit all circumstances.
- Many theories assert that leaders can change behavior to fit circumstances at will. However, many find it hard to do in practice, due to unconscious beliefs, fears or ingrained habits. Thus, he argued, leaders need to work on their inner psychology.
- None of the older theories successfully address the challenge of developing "leadership presence," that "certain something" in leaders that commands attention, inspires people, wins their trust, and makes followers want to work with them.

Leadership of Process Safety

As noted above, process safety leadership differs from general leadership only in focus. But leaders have struggled to include process safety in their focus. Stricoff (Ref 3.14) stated:

"The connection between leadership and process safety has not always been clear. Leaders often struggle to identify how or whether they affect process safety outcomes. The head of Transocean, for example, recently testified that while he wished his crew had done more to prevent the 2010 Deepwater Horizon disaster, his organization had found no failure of management. To many leaders, the idea that some events will 'just happen' despite leadership efforts is (and should be) deeply troubling.

"New research is showing that leaders play a critical and very specific role in catastrophic event prevention through their effect on culture. Of the 10 most recent events investigated by the U.S. Chemical Safety Board, each had at least one of four cultural factors as a root cause alongside [regulatory compliance] failures. A [PSMS] must be accompanied by a strong culture that requires critical leadership behaviors. If process safety leadership were a job description, there would be four basic competencies essential to success."

Leaders having these four competencies should:

- Have the conviction to lead safety,
- Understand how process safety works,
- Possess (and practice) great leadership skills, and
- Be able to influence people.

Motivated by culture lessons-learned from the 2005 incidents in Buncefield, Hertfordshire, UK and Texas City, TX, USA, the UK HSE in 2006 established a partnership of industry and regulators called the Process Safety Leadership Group (PSLG). The PSLG's goal is to drive high standards in process safety leadership in the UK and to implement recommendations made by the Buncefield Major Incident Investigation Board. PSLG (Ref 3.15) endorsed the competencies noted by Stricoff and recommended the following leadership actions:

- Address process safety leadership and culture at the Board of Directors' level, and include at least one Board member who is fully conversant in process safety to support the board's governance and strategic decisions,
- Engage the workforce in the development, promotion, and achievement of process safety goals,
- Provide sufficient resources at the operating and leadership level, all having the appropriate level of process safety experience; and
- Monitor process safety performance based on process safety leading and lagging indicators.

In other words, process safety leadership starts with the Board of Directors and senior leadership, and involves the entire organization. Everyone from the Board to the plant floor must have the necessary competence in process safety. The PSLG recognized that some Directors, especially those coming from other sectors, would not have competence in process safety. To compensate, they recommended that one Board member be highly experienced. See sections 5.3, 5.12, and Appendix D for more about competence.

After considering the leadership literature, Stricoff, PSLG, the expertise of the CCPS Culture Committee (See page xix), and other sources, some broad themes of process safety leadership can be seen:

- Achieve a balance between management and leadership:
    - o Establish clear roles and responsibilities for managers and others who function as leaders.
    - o Use management to define clear process safety work processes and manage them,
    - o Manage organizational change; and
    - o Use process safety metrics for decision making and balanced scorecards.

- Inspire subordinates and peers:
    - o Display visible support through felt leadership, leading by passionate example,
    - o Provide adequate, competent resources and annual budget for process safety; and
    - o Follow through on verbal support with personal actions.

These characteristics are expanded and described in much more detail in Sections 3.2 and 5.1, as well as in Appendix D.

3.2 CHARACTERISTICS OF LEADERSHIP AND MANAGEMENT IN PROCESS SAFETY CULTURE

Several key characteristics emerge from the Section 3.1 discussion of the basic themes of leadership in general and process safety leadership in particular. Like other core principals of process safety, the principal *Provide Strong Leadership* overlaps with other principals. Where appropriate in the following discussion, the overlap is noted.

Set the Tone

First and foremost, strong leaders/managers should set the overall process safety tone for the workplace. When leaders say, "Nothing is more important than safety," they should mean it. They should say it with a sense of vulnerability, as with the understanding of everything that process safety requires. This will help everyone else believe that the senior management believes fully in the importance of process safety. Without this belief, little else will be possible. Only the senior managers can establish this belief and it must be created in both word and deed. It not only starts with management/leadership, but it continues with them as well.

A single verbal message without follow-up actions, or no sustained transmission of messages will erode this belief. Also, inappropriate workplace behavior such as harassing behavior of any kind, unequal treatment by supervision or management such as favoritism or nepotism, or any other behavior that does not value and respect the people in the organization should not be tolerated in any way. It does not matter whether the behavior is face-to-face or occurs online. The existence of this type of workplace is a key cultural warning sign (Ref 3.16) of potential catastrophic incidents.

Leaders should understand that process safety culture is fragile. It can go from good to bad relatively quickly, but it will take a lot longer to reverse that trend. It takes years to create a positive PS culture, but only a few errant minutes to decimate it.

Influence and Inspire

Great process safety leaders have the ability to positively influence their subordinates', peers' and even superiors' behavior and work practices and inspire them to do the right thing. They earn the respect of those that they lead by both word and deed and this respect is not based on fear. Those with responsibilities in the PSMS will work hard to avoid disappointing leaders they respect. Although others may exhibit leadership behaviors and assume a leadership role in the PSMS, facility managers are in fact leaders of the PSMS. Therefore, more than others, facility managers should strive to be great process safety leaders. Good leaders serve as role models to the subordinates or their colleagues. To be most effective as PS leaders, people need to be capable of influencing up, down, and across the organization.

Act as Change Agents

Process safety leaders should be change agents, developing strategy for developing the process safety culture, and then improving and sustaining it. When selected, they volunteer, or they make themselves known simply by setting a positive example with their conduct. They should know the organization's process safety culture strategy and the rationale behind it. They should be the primary communicators of the strategy, shepherding it through the organization to successfully implement and sustain it (Ref 3.17).

Perhaps the most important leadership skill is to make the business case for process safety. This entails understanding risk at the operations level and being proficient at communicating that risk to upper management and garnering support for programs that reduce risk to the company's tolerance level.

Communicate effectively

Process safety leaders should be effective communicators of process safety program vision, goals, and objectives. They should communicate to their subordinates and colleagues frequently about process safety. Workers consistently indicate that the volume of communication about process safety compared to other topics greatly influences their perception of the importance of process safety versus other priorities. Process safety leaders who regularly and effectively communicate about process safety tend to make improvements (Ref 3.17).

Get on the Same Page

A poor relationship between organization work groups or between labor and management will never foster the kind of culture needed for safety or process safety to be successful. Additionally, process safety should not be a lever in labor negotiations. Rather, both sides should have arrived at a mutual understanding of the site's process safety vision through workforce involvement well in advance of negotiations. Process safety excellence is in the mutual benefit of both labor and management.

Nurture More Leaders

Organizations are dynamic, so leaders should nurture future leaders to succeed them to maintain a positive process safety culture going forward. Additionally, process safety leadership can and should exist across the organization in ways that cannot be expressed on an organization chart. Anyone at any level in an organization can exert strong process safety leadership capabilities and this should be encouraged and nurtured when it appears.

Say It with Feeling

Great process safety leaders demonstrate a passion for process safety that comes from an emotional commitment. More specifically, leaders should convey a deep respect for process hazards and an overall concern for peoples' safety. This may come from having experienced a past event or simply recognizing potential consequences if everyone does not fulfill their process safety roles. Expressing fear should be avoided, as this can lead to fatalistic attitudes or paralysis. Ultimately, it should be clear to all around that process safety is vitally important to the leader.

Some senior leaders achieve this by regularly communicating their greatest process safety concern, often expressed as "What keeps me up nights." This helps makes process safety personal. It may be easy and convenient to say it with feeling after an incident or significant near miss, but this is not sufficient. To be effective, the message must be delivered consistently over time.

Own Process Safety

Great process safety leaders accept total accountability for the PSMS and processes that they steward. They never point fingers at subordinates or others when things go wrong. Instead, leaders who own process safety find out where their leadership had been weak, and seek to improve.

Overall ownership for management and stewardship of PSMS and elements should be established at the highest levels of corporate and site management. Just as described by Rickover's Rules (Ref 3.18) for leadership in the nuclear navy, Process safety goals should be owned and driven by the operations and business. The process safety organization should be a valued resource to the leader, but the leader is ultimately accountable.

Establish Risk Criteria and Live by Them

Great process safety leaders accept total responsibility for the residual risk of their operations. Leaders should establish risk criteria and actions the organization should take if a risk does not meet the risk criteria. This may include establishing a risk level above which will not be accepted, a risk zone where risk should be reduced, and a risk level below which risk will be accepted. This responsibility falls solely on the organization's leadership and should not be delegated.

Leaders should encourage risk analysts to conduct their analyses thoroughly. While leaders may challenge conservatism, they should avoid even the appearance of influencing the outcome of risk analyses. Regardless of the outcome of risk analyses, leaders should address the results according to the established risk criteria.

Establish and reinforce stop-work authority

Leaders should make it clear that any employee can stop work or shut down the process if they perceive a potentially unsafe situation. Employees who exercise stop work authority should be complimented, not criticized. When stop work authority is used, leaders should avoid second-guessing the decision. Instead, understand the reason for the decision to stop work and address the root cause.

Create a Way of Being, Not a Program

Process safety culture is not a program or a distinct function or activity for which a person or group can be assigned responsibility or accountability. The existence of a process safety culture manager or coordinator likely indicates that the organization's leaders do not fully understand what process safety culture means. Only the senior leader can align the leadership team to achieve the corporate vision of process safety excellence. To be most effective, the PSMS must be fully integrated into the way business is conducted, not used as an occasional touchpoint or box to check.

Ensure Technical Competence

Individuals at all levels having process safety responsibilities should be technically competent in the relevant process technology, the specific process safety competence required for their job, and the PSMS in general. All such individuals should know the hazards of their process, the critical safeguards required to operate the process within the organization's risk tolerance, and their responsibilities in maintaining those safeguards. Process safety specialists should be able to accurately interpret how process safety regulations and other formal requirements apply to the facility's operations.

Technical competence is a hallmark of high reliability organizations (Appendix D), and gaps in competence are important warning signs of a potential catastrophic incident. (Ref 3.16). Technical competence will be discussed further in section 5.2.

Ensure Management Competence

In addition to possessing technical competence as noted above, managers should know how to manage. Management skills do not come automatically upon promotion to a management role. While an individual's personality traits may help them succeed in a management role, specific management skills must be taught, learned, and practiced. Management skills particularly relevant to process safety include:

- Knowledge of management systems, particularly the company's PSMS,
- Ensuring that employees (and managers themselves) operate within the constraints defined by the PSMS and the operating and maintenance procedures,
- Attention to detail, particularly of maintaining safeguards in full working order and approve all safeguard bypasses according to the corporate policy; and
- Verification of the PSMS performance within their scope of control

Candidate managers' competence related to the PSMS should be screened before their appointment. Any additional training or coaching needed should be identified and provided. While useful for all competencies, a thoughtful and up-to-date succession plan (See section 3.5) and organizational management of change procedure (OMOC) is especially helpful for process safety. Poor management skills are a key cultural warning sign (Ref 3.16) of potential catastrophic incidents.

Be Visible

Leaders should be visible in the field to evaluate conditions, understand site specific process safety issues, and be available to answer questions. Leaders should communicate process safety issues and requirements to site personnel in person, and seek productive feedback. They should engage the organization and assess if the line organization understands their responsibilities and performance expectations. Leaders should make process safety expectations and evaluations visible and explicit in their team members' individual goals and performance reviews.

Drive Good Morale, Especially During Change

Morale influences culture. Many of the things that drive good morale, such as trust, open communication, and a common imperative, also help drive good process safety culture. Conversely, poor morale often accompanies a poor safety culture. Good morale means more than general happiness in an organization. Additionally, a sense of pride and satisfaction permeates the organization, creating a sense of wanting to belong and of not wanting to disappoint their peers and leaders.

An old naval maxim says that while the ship's Captain delegates responsibility for nearly every duty to others, the morale of the crew is the one responsibility that cannot be delegated. The same holds true in industrial facilities. Morale is the responsibility of senior leadership and cannot be delegated.

Morale can be particularly hard to maintain during organizational changes such as new management, downsizing and changes in ownership. In addition to learning new roles, policies, and procedures, workers must deal with uncertainty about what the future holds for them personally. Leaders, especially new leaders, should pay particular attention to morale during such changes.

Understand Process Safety vs. Occupational Safety

Research at the Wharton School (Ref 3.19) revealed that there is no statistical correlation between the rates of process safety incidents and the rates of occupational injuries and illnesses. This can be easily understood by considering that occupational safety deals with how workers are protected, while process safety deals with how processes and equipment are designed, operated, and maintained.

Many leaders still do not understand this, despite numerous recent incidents that highlighted the issue. Process safety leaders should resist the temptation to infer that good occupational safety results mean that the PSMS is functioning well. Leaders should discuss indicators of process safety performance separately from occupational safety, treating each with equal importance and considering their unique differences.

Likewise, external recognitions of good safety performance should be considered carefully before assuming they address process safety. If a facility has earned the prestigious OHSAS 18001 certification, its safety management system may address process safety, but often it does not. Likewise, a facility that earns Voluntary Protection Program (VPP) Star status from US OSHA should be justly proud. However, VPP has historically focused much more on occupational safety than on process safety, and in recent years several VPP sites have experienced serious process safety incidents.

Use Metrics Prudently

The absence of process safety incidents and near misses does not necessarily mean that all is well, for two reasons. First, process safety incidents are rare by nature, and facilities can go many years without incident even as the conditions for an incident grow more and more likely. Second, the apparent absence of incidents may only be an indicator that incidents and near misses are not being reported. Even favorable results on leading indicators could be misleading if they are the result of "check-the-box" behavior, which can occur when management values the metrics over actual process safety performance.

Leaders should look behind the metrics. If lagging and leading indicators of the health of the PSMS are always positive and no problems are being identified, this could be an indicator of check-the-box mentality and should at least initially be a cause for concern. First verify that the metrics represent actual good performance in the field. If good performance is achieved, celebrate the teamwork and technical performance that achieved it, not the metrics themselves. Celebrating the metrics could foster check-the-box thinking by inadvertently implying that management values the metrics over the performance.

Use Monetary Incentives with Caution

In line with the above caution regarding metrics, leaders should exercise great caution when considering monetary incentives for achieving process safety related key performance indicators. Consider whether the incentive might foster check-the-box thinking or contradict the core principles of process safety culture in some other way and therefore backfire. See section 4.2 for a discussion of process safety culture and compensation.

No Fines does not Mean Strong Process Safety Performance

The absence of violations or findings from recent regulatory inspections does not guarantee that the PSMS is functioning as it should. Inspectors and auditors cannot know a facility's technology and management system as well as facility personnel, they visit the site only occasionally. Moreover, many process safety hazards do not fall under process safety regulations, but still must be managed to protect the company's assets, workers, neighbors, and reputation. Facilities should certainly seek to assure compliance, but gaps in the PSMS can exist even with the absence of citations and findings.

Reconcile Culture and Budget

Leaders should provide adequate financial and personnel resources to manage the process safety hazards of the facility. While leaders should challenge their teams to deliver results with efficient use of resources, cutting blindly will eventually erode a good culture. Instead, strengthening the process safety culture will help apply resources more efficiently. Conversely, throwing more resources at a weak process safety program cannot fix it in the presence of a poor culture.

Do Not Just Check-the-Box

Cursory MOC activities, rushed PHAs and PSSRs, and checking-the-box instead of completely performing the task are warning signs of a negative culture (Ref 3.16). Leaders should focus on the outcomes of the required tasks, not on completing the activity. In a resource- and time- constrained environment, this can be challenging. However, tolerating "checking-the-box mindset" is a big step towards normalization of deviance.

Do Not Blame

The vast majority of findings from process safety audits, incident investigations, and similar activities should be focused on management systems, process design, and technology. Generally, punitive actions should not be the outcomes of such activities, particularly those that measure or evaluate the health of the PSMS. An atmosphere of blame defeats trust and missed opportunities to learn and improve the culture.

There are rare exceptions where blame could be assessed. These include illegal acts and violating conditions of employment. However, even then, leaders should ask what gaps in the management system allowed the illegal act or violation to occur.

Encourage Bad News

Leaders should encourage the reporting of bad news. This helps assure that problems can be quickly surfaced and addressed. Employees should receive positive feedback when they report problems and deteriorating conditions, take precautionary action such as emergency shutdown, and suggest ideas for improvement. This should occur without fear of reprisal from management or from peers. In a well-led PSMS, the messenger is never punished, but instead encouraged. This promotes trust and open communication.

Do Not Allow "Not-Invented-Here"

Leaders should encourage improvement ideas the same way they encourage bad news. The not-invented-here syndrome has no place in a strong culture. Good ideas for process safety can come from anywhere: from any employee, any other unit or company site, other companies, and even from regulators and Non-Governmental Organizations (NGOs). Recognizing ideas that really make a difference is good, but in a strong culture, all ideas should be recognized whether used or not.

Trust, but Verify

In 1987, USA President Ronald Reagan and USSR General Secretary Mikhail Gorbachev signed an arms-reduction treaty based on mutual trust, enhanced by verification. While not everything in the PSMS needs to be verified, a sampling of them should be. Management should decide which PSMS activities deserve such checking, and then devise an independent and documented way of achieving the verification.

Candidates for verification may include:

- Closure of action items from audits, incident investigations, PHAs/HIRAs, MOCs and PSSR, emergency drill critiques, etc.,
- Lagging and leading metrics; and
- Training, i.e. that trainees achieve their learning goals.

Coordinate and Collaborate

Process safety has many diverse elements representing a wide range of functions and competencies. While the PSMS is intended to closely integrate these functions, few companies have achieved complete integration in practice. This is because many competencies, such as mechanical integrity, overlap with other broad functions at the plant, and it is advantageous to keep the functions together in organizations by specialty.

For this reason, PSMS managers often have no direct control over all the elements. The manager may have an indirect relationship with some functions, while others may operate totally independently. Facility leaders should create a culture where collaboration and coordination break down the silos, so they can ensure that all skill areas fulfill their process safety responsibilities and foster better integration of PSMS elements.

Avoid the "Flavor-of-the-Month"

As will be described in Section 3.4 the consistency of the process safety message is important. Competing and changing values make them seem temporary, rather than core to the organization, while changing goals may come to be considered optional. Leaders should deploy goals thoughtfully, strategically and systematically. The Organization for Economic Cooperation and Development (OECD) summarizes this succinctly (Ref 3.20):

"The CEO and other leaders create an open environment where they:

- *Keep process safety on their agenda, prioritise it strongly and remain mindful of what can go wrong,*
- *Encourage people to raise process safety concerns or bad news to be addressed,*
- *Take every opportunity to be role models, promoting and discussing process safety,*
- *Delegate appropriate process safety duties to competent personnel whilst maintaining overall responsibility and accountability,*
- *Are visibly present in their businesses and at their sites, asking appropriate questions and constantly challenging. the organisation to find areas of weakness and opportun-*

ities for continuous improvement; and
- *Promote a safety culture that is known and accepted throughout the enterprise."*

3.3 LEADERSHIP VS. MANAGEMENT

Leadership means more than simply title and the assigned authority that comes with it. Although today, managers tend to be expected to be leaders, leadership and management differ significantly in the competencies exercised. Many companies use the term Leadership Team to describe the facility manager and their direct reports. Although this term expresses the hope and expectation that these reports will exhibit strong leadership, sometimes this group act as managers rather than leaders.

Success in leading process safety culture depends on being both a manager and a leader. Typical differences between leadership and management are summarized in Figure 3.1.

Figure 3.1 Differences between Leader and Manager

Leader	Manager
Creates and communicates a future vision	Develops plans and allocates resources
Encourages others to commit to the vision	Sets objectives and organizes schedules
Sets direction, and creates alignment	Directs and monitors
Motivates and inspires	Creates order and efficiency
Helps organization develop and adapt	Ensures standards are met
Creates new structures	Leverages current structures

(After Refs 3.21, 3.22)

Strong process safety leadership refers to the ability of a person to convince his/her reports and peers of the right process safety thoughts and actions – winning their hearts as well as their minds. Senior managers should be process safety leaders. Additionally, in a strong process safety culture, mid-level managers, supervisors, technical specialist, and even front-line employees can and should be leaders also. True leadership is not conveyed by one's position on an organization chart.

Effective leaders inspire their reports and co-workers and earn their respect with direction and advice that is sound and consistent. Leaders accept direct accountability for all things that occur within their sphere of responsibility. They do not attempt to publicly place blame on their subordinates when things go wrong.

More than anything, subordinates will not want to disappoint someone who has earned their respect as a leader. Visionary and inspiring managers who are also good leaders are committed to doing what is right, and demonstrate their values through their communications, actions, priorities, and provision of resources.

3.4 CONSISTENCY OF PROCESS SAFETY MESSAGES

In its investigation of process safety culture in BP's USA refineries, the Baker Panel (Ref 3.23) found that workers had received many messages from management over the years addressing many values. These tended to dilute the importance of any value generally, and certainly of process safety.

This happens in many companies. Leadership communicates mission and vision statements, core values, central tenets, and overarching principles, wishing to better define what their organizations stand for and how they operate. Many times, these supplement company codes of conduct and address some of the same issues. Over time, these messages shift with changes to management, ownership, and business climate.

Ideally, these communications should be aligned and flow logically. However, sometimes messages can be competing, confusing, self-diluting, or so unrealistic that they are dismissed offhand. Issuing or reissuing these communications using currently trendy management vocabulary may look good in marketing but completely confuse workers

Aspirational goals, mission and vision statements, and the like can be a positive step, but only if they are:

- Concise and easy to understand,
- Coherent, with vision flowing logically to mission, strategy, and goals,
- Consistent and frequently reinforced; and most importantly; and
- Followed with positive action.

Additionally, the various statements of values as they apply to process safety should remain consistent with time, not changing as business environment and other corporate priorities shift. Process safety management systems and the underlying culture should be part of the value system of the organization. Even if the language changes, the meaning and values should be consistent over the long term.

3.5 TURNOVER OF LEADERSHIP, SUCCESSION PLANNING, AND ORGANIZATIONAL MANAGEMENT OF CHANGE

Stability in facility leadership positions can be important to process safety performance in several ways, as discussed in the following paragraphs.

Turnover of Leadership

The Baker Panel (Ref 3.23) noted:

"Most of BP's five U.S. refineries have had high turnover of refinery plant managers, and process safety leadership appears to have suffered as a result."

A site having both a strong PSMS and strong culture may be able to tolerate some turnover at the senior positions without materially affecting process safety culture, but only if new managers embrace and support the existing culture. Towards this end, some companies require or suggest that new facility managers refrain from changing visions, policies, and organizational structure until they have been in their roles for a specified time.

In an environment of high turnover, weak PSMS, and weak culture, managing process risks may depend disproportionately on the capabilities, efforts, and even personalities of individuals. These individuals' jobs become harder by real or perceived shifting priorities from leader to leader. And if one or more of the individuals leave their role, the system can become gravely compromised.

Companies working to strengthen process safety culture and management systems should consider maintaining a stable leadership structure during the process. Frequent change of facility manager can make it harder to establish with the workforce a shared vision about process safety as a core value. It can take time to build trust. Frequent turnover may not allow sufficient time, and may demotivate either the leader or workers to make the effort to build trust, leading to possible alienation. Experience shows that 3-5 years of stability is usually required.

Succession Planning

Succession planning is an important way to maintain process safety culture and PSMS performance through leadership changes. A good succession plan, supported by a sound organizational management of change process, helps maintain competency, performance, and culture during organizational changes.

A succession plan ensures that qualified and motivated employees are ready to take over when a key person leaves the organization. Whether or not the actual successors are known, a succession plan includes experience and competency requirements for potential replacements. Having a succession plan demonstrates to stakeholders that the organization is committed maintaining consistent functioning at all times, including during times of transition. The HR Council (Ref 3.24) offered an example highlighting what can happen without succession planning:

A mid-sized organization relied heavily on the corporate memory, skills and experience of a longtime employee. In her final position, she was responsible for office administration including payroll and budget monitoring. During her career, she held many positions and understood well the organization's operations and history. Her unexpected death was both an emotional blow and a wake-up call to her colleagues. Everything she had known about her job was "in her head." While management discussed regularly the need to document her knowledge to pass it on to others, this had never happened. Ultimately, the organization did regroup and survive the transition, but employees experienced high stress as they struggled to determine what needed to happen when. A great deal of time and effort was spent recreating systems and processes and even then, some things fell through the cracks resulting in the need to rebuild relationships with supporters.

The organization in this example did not have process safety or operational risks, but if they did, imagine the potential impact when "some things fell through the cracks."

Most companies realize the value of succession planning and some have formal succession plans. One strong proponent of succession planning was former General Electric Company CEO Jack Welch. During his tenure as CEO, Welch led the transformation of GE and increased its shareholder value many-fold. However, he knew that GE had to groom his successor as well as successors for other key employees. Welch knew that succession planning avoids disruption and employee trauma when key personnel change, whether the departure is anticipated or not. Succession planning should be company policy, dealt with openly and deliberately by corporate boards (Ref 3.22).

From a process safety culture standpoint, two key points stand out. First, the succession planning for future leaders should include preparing successors to assume their executive process safety roles. Second, all process safety critical positions should be considered for inclusion in the succession planning process at some level. A study conducted by Bersin & Associates (Ref 3.22) found that most companies implement succession planning at only the most senior executive levels. Fewer than 40 percent of the respondents said their companies included midlevel managers and skilled professionals in succession planning initiatives, and just 11 percent included first-line supervisors. The study concluded that "enduring organizations" – those that survive and prosper a long time – execute succession management practices across all levels of the organization.

The Bersin study highlights some additional benefits of succession planning:

- Alignment between the organization's vision and the human resources strategic staffing function,

- Enhanced recruiting and retention by demonstrating a commitment to developing career paths and advancement opportunities; and
- Telling employees that they are valuable.

Notwithstanding the preceding discussion, the complete absence of turnover should also be avoided. If a leader – especially the most senior leader – holds the same position in an organization for a very long time, a type of cult mentality can develop. In this situation, the entire organization takes on the personality of the leader and the identity of the organization and the leader becomes blurred. Effectively this removes empowerment and trust from the culture, because every action flows from the leader. Moreover, in such a situation, no matter how strong the culture, the eventual succession of the leader may be very difficult, because the successor will typically have new ideas and a different leadership style. Also, the longer a leader stays in a given position, the less objective they can become. They can use the "we've tried that before' excuse to deflect necessary improvements or innovative methods

Organizational Management of Change

In recent years, many companies have voluntarily adopted Organizational Management of Change (OMOC), either as a separate PSMS element or as part of their existing MOC. Formal OMOC processes allow an organization to formally examine the impacts of planned or unplanned turnover of leaders and

employees with process safety critical roles. This usually includes key managers, leaders, technical experts, and skilled positions.

As explained in section 3.2, some process safety leaders may not be associated with a key role, but may instead evolved purely because of the leadership characteristics they display. Companies should consider whether to include such individuals with the OMOC process.

Significant guidance for OMOC has been published by CCPS (Ref 3.25). CCPS also notes that establishing a consistent and sustainable PSMS and culture across the enterprise can greatly help both succession planning and OMOC. This helps leaders moving between units, facilities, and businesses develop and train to qualify for their new roles. They can then start their new roles with less disruption to the management system and culture. The more comprehensive and corporate-wide the management system, the easier it will be to manage organizational change.

3.6 SUMMARY

Leadership and the culture are inextricably linked. Indeed, the presence of strong positive and felt leadership most strongly influences the nature of the process safety culture. As a foundational core principle of the process safety culture, leadership enables the other core principles to be established and nurtured. Without the example of strong leadership, both by word and deed, the process safety culture will suffer, as will the PSMS.

Several factors help leaders set the right tone, including technical competence of managers, low or stable turnover rates for key leaders, a well-designed succession plan, and effective OMOC to help ensure continuity.

As E.H. Schein (Ref 3.26) said:

"Culture and leadership are two sides of the same coin in that leaders first create cultures when they create groups and organizations ... The bottom-line for leaders is if they do not become conscious of the cultures in which they are imbedded, those cultures will manage them. Cultural understanding is

desirable for all of us, but it is essential to leaders if they are to lead."

3.7 REFERENCES

3.1 Chemers M., *An integrative theory of leadership*. Lawrence Erlbaum Associates, Publishers, 1997.

3.2 Chin, R., *Examining teamwork and leadership in the fields of public administration, leadership, and management*, Team Performance Management, 2015.

3.3 Markwell, D, *Instincts to Lead: On Leadership, Peace, and Education*, Connor Court, Australia, 2013.

3.4 Bird, C., *Social Psychology*. New York: Appleton-Century, 1940.

3.5 Stogdill, R., *Personal factors associated with leadership: A survey of the literature*, Journal of Psychology, Vol. 25, 1948.

3.6 Mann, R., *A review of the relationship between personality and performance in small groups*. Psychological Bulletin, Vol. 56, 1959.

3.7 Arvey, R., Rotundo, M., Johnson, W., Zhang, Z., & McGue, M., *The determinants of leadership role occupancy: Genetic and personality factors*. The Leadership Quarterly, Vol. 17, 2006.

3.8 Judge, T., Bono, J., Ilies, R., & Gerhardt, M., *Personality and leadership: A qualitative and quantitative review*. Journal of Applied Psychology, Vol. 87, 2002.

3.9 Tagger, S., Hackett, R., Saha, S., *Leadership emergence in autonomous work teams: Antecedents and outcomes*, Personnel Psychology, Vol. 52, http://onlinelibrary.wiley.com/doi/10.1111/j.1744-6570.1999.tb00184.x/abstract, 1999.

3.10 Kickul, J., Neuman, G., *Emergence leadership behaviors: The function of personality and cognitive ability in determining teamwork performance and KSAs*, Journal of Business and Psychology", Vol. 15, 2000.

3.11 Smith, J., & Foti, R., *A pattern approach to the study of leader emergence*, The Leadership Quarterly, Vol. 9, 1998.

3.12 Foti, R., & Hauenstein, N., *Pattern and variable approaches in leadership emergence and effectiveness*, Journal of Applied Psychology, Vol. 92, 2007.

3.13 Scouller, J. (2011). *The Three Levels of Leadership: How to Develop Your Leadership Presence, Knowhow and Skill.* Cirencester: Management Books 2000.

3.14 Stricoff, S., *What Process Safety Needs in a Leader*, Safety + Health, 2013.

3.15 United Kingdom Health and Safety Executive (HSE), *Safety and environmental standards for fuel storage sites*, Process Safety Leadership Group, Final report, 2009.

3.16 CCPS, *Recognizing Catastrophic Incident Warning Signs in the Process Industries,* American Institute of Chemical Engineers, New York, 2012.

3.17 Mathis, T., Galloway, S., *STEPS to Safety Culture Excellence^SM*, Wiley, 2013.

3.18 Paradies, M., *Has Process Safety Management Missed the Boat?* AIChE, Process Safety Progress, Vol. 30, No. 4, 2011.

3.19 Elliot, M., et. al., *Linking OII and RMP data: does everyday safety prevent catastrophic loss?* International. Journal of Risk Assessment and Management, Vol. 10, Nos. 1/2, 2008.

3.20 Organization for Economic Cooperation and Development (OECD), Corporate Governance for Process Safety, OECD Guidance for Senior Leaders in High Hazard Industries, June 2012.

3.21 United Kingdom Health and Safety Executive (HSE) Health & Safety Laboratory, *Safety Culture: A review of the literature*, HSL/2002/25, 2002.

3.22 Oracle, *An Oracle White Paper, Seven Steps for Effective Leadership Development*, June 2012.

3.23 Baker, J.A. et al., *The Report of BP U.S. Refineries Independent Safety Review Panel*, January 2007 (Baker Panel Report).

3.24 HR Council, *HR Planning, Succession Planning* (http://hrcouncil.ca/hr-toolkit/planning-succession.cfm)

3.25 CCPS, *Guidelines for Managing Process Safety Risks During Organizational Change,* American Institute of Chemical Engineers, New York, 2013.

3.26 Schein, E.H., *Organizational Culture and Leadership*, 3rd Ed., San Francisco: Jossey-Bass, 2004.

4

APPLYING THE CORE PRINCIPLES OF PROCESS SAFETY CULTURE

This chapter discusses the application of the core principles of process safety culture described in Chapter 2 and the leadership principles of process safety culture described in Chapter 3, including:

- Human behavior,
- Ethics,
- Compensation,
- External influences; and
- Metrics.

4.1 HUMAN BEHAVIOR AND PROCESS SAFETY CULTURE

Every PSMS is performance based, in one way or another. Corporate risk criteria are determined, either by the company, by society through regulations, or in some cases a trade association. Then the PSMS operates to control risk to meet these criteria. In other words, PSMSs are performance-based. Whose performance? Regardless of how much software is used to facilitate a PSMS, ultimately it is run by people who must interact. This puts human behavior and culture in the center of every PSMS.

Essential Practices for Creating, Strengthening, and Sustaining Process Safety Culture, First Edition. CCPS, © 2018 AIChE. Published 2018 by John Wiley & Sons, Inc.

Human behavior usually contributes to process safety incidents through human errors. This includes acts of omission – something a person fails to do – and acts of commission – something a person does that they should not. Experience and research has shown that some human error is inevitable. But human error can be affected by so-called performance-shaping factors, stresses and influences that increase or decrease human error. Many of the performance-shaping factors can be managed through the practice of human factors design, discussed in depth by CCPS (4.1). Culture also plays a significant role.

A person could rightfully ask, "Does behavior create the culture, or does the culture create the behavior?" Arguments can be made either way. A key premise of this book has been that the behavior of people engaged in any set of tasks will be affected by the culture surrounding those tasks. However, the culture itself can be strengthened or weakened by behaviors. Indeed, behaviors aligned to the culture core principles discussed in this book should strengthen culture. Meanwhile, behaviors counter to the culture core principles, such as breaking trust or normalization of deviance, weaken culture.

The research of Daniellou (Ref 4.2) supports this notion of human characteristics that can be influenced by culture to drive behavior. These characteristics can be driven by positive culture to create good safety behavior, and can be driven by negative culture to threaten safety.

Daniellou noted that human errors, though generally unintentional, are made as the result of conscious acts performed without malice. That is, people choose to perform incorrectly, and do so with good or neutral intentions. In this context, associating error with words such as "fault" or "liable" is doubly counter-productive. Not only does this prevent the organization from identifying the real reason, it also prevents *open and frank communication* and destroys *mutual trust*. Only errors where there was some malice or gross negligence involved should be the subject of disciplinary action. Daniellou's work on human behavior and industrial culture is described in more detail in Appendix G.

In other words, each of the 10 culture core principles described in this book can help drive improved process safety culture and improved performance at a facility if implemented sincerely. Conversely, if omitted or implemented insincerely, culture and performance can be profoundly harmed. It is not possible to separate the process safety culture from the behavior of the human beings that develop, administer, and execute the program. All PSMS dimensions and behaviors, from determining risk criteria to validating the tasks done to operate within the risk criteria, will be profoundly affected by the process safety culture of the facility and, if applicable, its parent company.

4.2 PROCESS SAFETY CULTURE AND COMPENSATION

Money is one of the strongest influences on human behavior, for better or worse. When developing compensation and incentive schemes based on process safety performance, it is critical to design them carefully to reinforce the desired cultural attributes and behaviors. It is equally critical to be aware of the many pitfalls that can lead a well-intentioned compensation scheme to unwittingly support negative behaviors.

These considerations are not very different from any other incentivized business goal that appears on a balanced scorecard. For example, if the company has a goal of top line growth, individuals should obviously be incentivized on their actual or team contribution to the top line. However, if their contributions are based on sales dollars, it might drive them to sell at an unprofitable price, or make deals where they ship extra product that can be returned for credit, driving up their individual sales figures at the expense of increased return costs.

In general, when designing an incentive system for process safety, the following points should be considered.

Consider the Potential for Inverse Effects

A company goal to reduce the number of incidents from year-to-year is certainly desirable. However, using incident number or incident reduction for purposes of incentive may drive personnel to hide or under-report incidents. Whatever basis for incentive is considered, leaders should think about how it could lead to the opposite of the desired behavior. It may also make sense to independently validate incentive metrics to ensure this has not happened.

Focus on the Frequent, not the Rare

Since the ultimate goal is to prevent process safety incidents, it can be tempting to use the lagging process safety incident rate as the basis for incentive. The problem is that incident rates are generally low, and a leader can perform poorly in process safety for a long time before an incident occurs. It is better to avoid using lagging metrics, such as incident rate. Instead use leading metrics related to correct behaviors that must happen frequently over time such as percent completion of asset integrity actions (e.g. inspection, testing, and preventative maintenance). Near-misses occur much more frequently and can also be an option.

Focus on the Long-Term, not the Short-Term

Since process safety needs to be performed well, consistently over time, the basis for incentives should consider long-term performance. This can be easier to accomplish in incentive schemes that have a multi-year basis, but still is possible in year-by-year schemes. For example, the incentive should consider whether the goal was intended to have been reached by steady performance over the year, and penalize individuals who achieved the desired result by focusing all their efforts in the final months. Companies could also consider paying out the incentive for the prior year over time, with the stipulation that the payout could be suspended if the desired behavior slips.

Balance Leading and Lagging Metrics for Incentives

Simply looking at reactive or lagging metrics can create a very narrow focus. It would be preferable to base incentives on leading metrics or activities that reinforce a positive PS. Activities such as MI inspections completed on-time, reducing backlog of high-risk PHA action items, or percentage of operating procedures reviewed within the specified frequency drive positive behaviors and should drive improved PSMS performance.

Design Group Incentives Carefully

Process safety culture is ultimately a group activity, so incentivizing each member of a group to collectively address a process safety goal can help the group rally to that goal. Peer pressure can exaggerate the positive aspects of incentives, with group members pushing each member to achieve the goal. It can also exaggerate the negative aspects, with some group members pushing others to hide incidents and normalize deviance. Finally, group incentives should not prevent the organization from also assigning individual goals.

Change the Basis of Incentives

In the long term, using the same basis for incentives long after the desired behavior has become normalized can lead to the incentives being considered quasi-permanent by the workers. When, after time, the organization changes the incentive basis, as it ultimately will, resentment and loss of *trust* could result. To avoid this, companies should change the basis of process safety incentives from time to time. Ideally the new basis should be based on improving an area of the PSMS not addressed in the recent past. However, take care not to change too frequently to avoid appearing to pursue the "flavor of the month."

Base Process Safety Incentives on Process Safety Metrics

In several chapters, this book has cautioned the reader about confusing occupational safety metrics and process safety metrics. The same care should be taken in the design of process safety incentives to ensure the incentive is actually measuring a key aspect of the PSMS or process safety culture. For the same reason, companies should not combine process safety indicators with occupational safety indicators into a single safety incentive.

Consider Process Safety as a Multiplier Instead of an Adder

Most balanced scorecards have additive contributions from many goal areas. This can allow a manager to choose to excel in one goal area and mostly ignore another. This behavior can be exaggerated in schemes where exceeding a goal can be rewarded with a higher multiplier, and can reward managers for prioritizing production over process safety. With the *imperative for process safety* in mind, companies can consider developing a process safety performance factor as a multiplier in calculating the entire incentive. In other words, if a manager's process safety performance is poor, their factor could be zero and they would get no incentive at all, regardless of how well they did in other areas.

Weigh Process Safety Incentives Consistently with Other Priorities

Companies may not choose to use the process safety multiplier approach mentioned immediately above. If not, to address the same concerns, the fraction of the incentive related to process safety should make it undesirable for a manger to ignore process safety in favor of other business areas.

<u>Should There be a Process Safety Incentive at All?</u>

Should the organization forego all process safety-related incentives because it is simply the right thing to do? Choudhry (Ref 4.3) argues that working without injury should be a strong incentive by itself, as it provides workers with the long-term term benefit of being able to provide earnings for the company and themselves and their families. However, money is a very strong human motivator, and if used with care can help change behavior. This decision may be influenced by where a company is on its culture improvement journey. Nonetheless, good process safety performance should be rewarded at the very least by a heartfelt thank you from the leadership team.

<u>Summary</u>

Incentives can be particularly useful in underlining the core principles *establish an imperative for safety* and *combat the normalization of deviance*. They have the potential to influence process safety performance and process safety culture, both positively and negatively. Ultimately, metrics and incentive approaches should treat process safety on par with other business priorities, discourage managers from prioritizing production over process safety, and drive the desired results and behaviors. Leaders should examine incentives schemes to make sure they do not drive the opposite or negative behaviors.

4.3 PROCESS SAFETY CULTURE AND ETHICS

Krause (Ref 4.4) links process safety and ethics closely:

"(Process) Safety appeals to the ethical ideals that motivate a company's best leaders at every level of responsibility."

Definition of Ethics

Ethics is defined as (Ref 4.5):

- The study of the general nature of morals and of the specific moral choices to be made by the individual in his relationship with others.
- The rules and standards governing the conduct of the members of a profession.
- Any set of moral principles or values.
- The moral quality of a course of action.

For the purposes of this book process safety culture was defined in Chapter 1 as:

"The pattern of shared written and unwritten attitudes and behavioral norms that positively influence how a facility or company collectively supports the successful execution and improvement of its PSMS, resulting in preventing process safety incidents."

Clearly, ethics and process safety culture are closely related concepts. They share an almost total reliance on how people feel about certain aspects of their jobs and how they behave. Each also shares a reliance on rules, standards, procedures, and managements systems. However, several things differentiate them. For example, morality is the basis for ethical behavior, while process safety culture is based on the human value of preserving life and property. However, like ethics, a positive safety culture does include moral behaviors that are fair, honest, and open.

An interesting question then emerges: does the process safety culture drive the ethical behavior of an organization or does the ethical behavior drive the culture? Clearly, good or bad ethical behavior of influential persons can affect the culture for good or bad. Likewise, good or bad culture can affect the ethics of an organization. However, it is hard to establish good ethics in the absence of a good culture.

Ethical behavior is marked by the ability to make the "hard call" when required. In extreme situations where the stakes are high, acting ethically requires great personal courage. Making decisions that run counter to competing pressures is one of the most demanding situations that people can face. When the competing pressure is economic, a decision-maker may feel that resisting that pressure would put their livelihood and ability to support their family at risk.

Other competing pressures include desire for personal gain, avoiding embarrassment, and desire to please a certain constituency. Any such competing pressure could influence a person to bend or violate ethical principles. People who withstand these pressures and make the right call are sometimes referred to as having a steady "moral compass." As McNamara describes (Ref 4.6), the moral principles or values that underlie ethical behavior and guide behavior can easily become compromised when people are placed in stressful situations Stress or confusion are not excuses for unethical behavior, but they are reasons. McNamara also asserts that there are two broad challenges to business ethics:

- **Managerial mischief:** Managerial mischief includes illegal, unethical, or questionable practices. There has been much written about managerial mischief, leading many to believe that business ethics is merely a matter of preaching the basics of right and wrong. More often, though, business ethics requires dealing with dilemmas that have no clear indication of what is right or wrong.
- **Moral mazes:** The other broad area of business ethics is "moral mazes of management." These include the numerous ethical problems that managers must deal with

daily, such as potential conflicts of interest, wrongful use of resources and mismanagement of contracts and agreements.

Characteristics of Ethical Organizations

Ethics experts have identified several characteristics common to highly ethical organizations: (Refs 4.6, 4.4):

- They are led by people committed to process safety and ethics for its own sake, not solely for compliance - *Establish an Imperative for Safety.*
- They see their activities in terms of a purpose that members of the organization highly value, which includes ethics and process safety. They also see leaders act consistently and credibly with that purpose. And purpose ties the organization to process safety. *Establish an Imperative for Safety, Provide Strong Leadership.*
- They are obsessed with fairness. Their ground rules emphasize that in any relationship, the other persons' interests count as much as their own. Workers see leaders acting fairly and know they can follow instructions without fear of mistreatment. *Foster Mutual Trust.*
- They keep communication channels open, especially upward. Leaders and supervisors then act responsively, even to bad news. This helps ethical issues to surface before they become a crisis. Leaders also encourage peers to communicate on sensitive issues including those of process safety and ethics and other critical areas. *Foster Mutual Trust, Defer to Expertise.*
- Responsibility is individual rather than collective. Each person assumes personal responsibility for their actions in support of the organization. The organizations' ground rules mandate that individuals are also responsible to themselves. *Provide Strong Leadership.*

- They are at ease interacting with diverse internal and external stakeholder groups. The ground rules of these firms make the good of these stakeholder groups part of the organizations' own good. *Ensure Open and Frank Communications, Provide Strong Leadership.*

Krause goes on to draw a strong connection between a safety program and the ethical behavior of an organization and concludes:

> *"This foundation in principle perhaps is the greatest strength that safety offers to organizations interested in ethics. Safety appeals to the ethical ideals that motivate a company's best leaders at every level of responsibility." (Ref 4.4)*

Within the context of process safety management programs, the characteristics espoused by McNamara and Krause are enabled by a positive process safety culture that allows them to flourish. The core principles of process safety culture involved in this enabling have been added to their statements for emphasis. See Chapter 5 for the effects of process safety culture on each element of a PSM/RBPS program. Also, there are several examples and important topics addressed in Chapter 2 that overlap with ethics or extend the process safety culture into behaviors and decision making regarding the PSM/RBPS program elements and their activities.

Ethical Dilemmas

When the topic of business ethics arises, people are quick to speak of fairness, honesty, and "doing the right thing." However, sometimes the right thing is not clear, leading to ethical dilemmas. These can exist when there is presence of a) significant value conflicts among differing interests, b) real alternatives that are equally justifiable, or c) significant consequences on stakeholders in the situation. When presented with complex ethical dilemmas, most people realize there's a wide gray area in applying ethical principles. Most ethical dilemmas faced by persons in the workplace are highly complex.

Also, most people want to avoid conflict and seek a "win-win" outcome. However, many decisions must be made in which some parties will be satisfied, while others will not. Some of these decisions are "black and white" where a clearly defined right or wrong path forward presents itself. However, many decisions are not as starkly defined and have various shades of gray.

Grubbe (Ref 4.7) points out ethical dilemmas in noting that "The underlying causes of an incident may have little or nothing to do with technology, for example:

- You do what your boss tells you to do, even if it is against your better engineering judgment.
- You tell your boss about a condition that could be dangerous under certain conditions. When your boss says that everything is fine, you remain silent and do not revisit the subject.
- You act contrary to a legal hold order and destroy evidence related to pending litigation, because you are afraid you will lose your job, or be prosecuted for something you wrote.
- You feel your job is in jeopardy because you disagree with your boss.
- You are asked to revise the language of a report to downplay or remove results that do not support the desired conclusions of the organization or certain senior personnel.
- You have been asked to alter data to strengthen the outcome or conclusions of research."

From the process safety perspective, these dilemmas can arise in the context of manufacturing operations, engineering design project budget and schedule, and process engineering. The individual could fear embarrassment, reprisal, or being labeled as a "non-team player," among other consequences. Of course, violating the provisions of a subpoena or other order of the court is a serious lapse of ethical behavior and is a criminal act for which the consequences could be very severe.

Allan McDonald (Ref 4.8), a Morton Thiokol engineer and director, faced a severe ethical dilemma before the scheduled launch of the Space Shuttle Challenger. Because the cold weather forecasted for the night before the launch would cool the boosters' O-rings well below the temperature they had been proven to seal effectively, McDonald refused to sign the launch consent form.

NASA then demanded that Thiokol senior management affirm McDonald's decision in writing. In the press of time and under significant pressure from NASA, a Thiokol senior executive at headquarters over-rode McDonald's decision and signed the consent form. Investigation revealed that the executive wanted to give his customer, NASA, what they wanted to keep their business. So, he bowed to what NASA refers to as go-fever.

McDonald displayed much ethical courage in taking a firm stand asserting the *imperative for (process) safety*. He did so, knowing that his resistance to NASA's pressure to proceed would likely cost him his job and brand him as a non-team player in the aerospace community. Indeed, after Columbia was destroyed and its seven-person crew lost, McDonald was demoted in the hope he would resign.

When the Rogers Commission investigating the accident, learned this, McDonald was asked to testify before the Commission and Congress. Following his testimony. Congress showed *strong leadership* in passing an act restoring McDonald to his previous position, the only time Congress ever directly intervened in the employment of a government contractor.

<u>Managing Ethics</u>

Ethical behavior is not an innate activity. Ethics can, and should, be managed. Many companies define standards guiding ethical behavior that is encouraged as well as unethical behavior that is not tolerated. Likewise, professions and professional societies, as well as trade organizations, have had similar codes for decades and even centuries. Codes generally represent high-level aspirations of conduct, and establish consistent expectations for the conduct of members of the group. Moreover, they declare to the public the behavioral expectations of the group.

Organizations associated with process safety that have ethical codes include the Board of Certified Safety Professionals, the Board of Environmental Auditor Certification, and the National Society of Professional Engineers, among others. The American Institute of Chemical Engineers (AIChE) stands out among technical societies for its Code of Ethics. The preamble to AIChE's code states (Ref 4.9):

"Members of the American Institute of Chemical Engineers shall uphold and advance the integrity, honor, and dignity of the engineering profession by: being honest and impartial and serving with fidelity their employers, their clients, and the public; striving to increase the competence and prestige of the engineering profession; and using their knowledge and skill for the enhancement of human welfare."

The AIChE code goes on to address three topics closely related to process safety culture:

- (1) Hold paramount the safety, health and welfare of the public and protect the environment in performance of their professional duties. (*Establish the imperative for process safety*),

- (2) Formally advise their employers or clients (and consider further disclosure, if warranted) if they perceive that a consequence of their duties will adversely affect the present or future health or safety of their colleagues or the public. (Ensure open and frank communications); and
- (11) Conduct themselves in a fair, honorable and respectful manner. (*Foster mutual trust*)

Over the years, laws and regulation have strengthened to enforce ethical conduct, typically defining ethical requirements in many business areas, deviance from which can be penalized. Specific to process safety, many countries' regulations increase the penalties for non-compliance when the violation is deemed to be the result of an unethical choice (In the USA, termed a willful violation). Like most new laws and regulations, their scope is hotly debated. Nonetheless, the general trend represents the public's intolerance of unethical behavior, especially if members of the public can be harmed personally or financially by that conduct.

Symbols and ceremonies can help remind people of their ethical obligations. One of the most famous of these is the practice begun in Canada in 1922. Since that time, all engineering graduates participate in the Ritual of the Calling, in which they are reminded of the ethical obligations of their future careers. Each engineer is presented with a ring originally made from the scrap iron reclaimed from a bridge that collapsed due to poor design. The engineers wear the ring on the little finger of their writing hand as a continuous reminder of what can happen if they fail in their engineering and ethical duties. This practice has since been copied in a limited way in the USA.

Companies wishing to unite employees around process safety culture and ethics could consider similar kinds of symbols and ceremonies. At least two companies include quasi-religious ceremonies at the beginning of each shift to remind workers and supervisors of their ethical obligations to safety, the environment, and each other. More traditionally, companies should remind all employees of their ethical requirements, procedures for reporting unethical behavior, and resources to use when in an ethical dilemma.

Many professional certifications and professional engineering licenses include questions about ethics in their qualifying exam or continuing development requirements. These requirements are not necessarily difficult to meet, but as above, serve as a ritual of passage and a continuing reminder of engineers' ongoing ethical obligations.

If a company's ethics policy or code of conduct does not fully address decisions related to process safety, that gap should be closed. The policy should then be supported by every element of the PSMS and embedded in the operations and engineering dimensions of process safety. Ideally, the company should also have an ethics hotline and process to support resolution of ethical dilemmas.

In conclusion, achieving strong corporate ethics requires leaders and managers to serve as ethical role models, set expectations, and enforce the ethics policy in every aspect of their daily work.

Process Safety Culture Core Principles & Ethics

Ethics are represented most strongly by 4 of the 10 process safety culture core principles

- *Establish an Imperative for Safety*: Is it safe to proceed?

No one should be placed in the position of having to prove that a situation is unsafe. Rather, those responsible for production should have to prove that it is safe to proceed. And like Allan McDonald's management at Morton Thiokol, decision makers should not be placed in an ethical dilemma pitting process safety against operations or schedule. This creates a significant ethical dilemma for those advocating for safety. Also see section 2.1.

- *Provide Strong Leadership*: Be a Role Model

Leaders and other decision makers who receive bad or undesired news should support the bearer of that news if they are technically right. Otherwise, those with the correct information will be placed in the ethical dilemma of whether to become a whistle blower in order to get the vital information to those who are authorized to stop the course of action decided.

- *Ensure Open and Frank Communications*: Be Direct and Welcome Bad News

Ethical (and cultural) expectations should be communicated directly and clearly. Channels of communication should be open and sufficiently streamlined that bad news can easily reach leadership. People in possession of information that needs to reach senior members of the organization should not be placed in the ethical dilemma of having to break the rules to deliver the information.

- *Foster Mutual Trust*: No Fear

When people must report bad or disappointing news, they should not be put in the ethical dilemma of fearing reprisal, isolation, or loss of their job if they do.

Ethics Within Process Safety Management System Elements

As a process safety management program is implemented and its daily activities are executed, situations may sometimes develop that place people carrying out those activities in ethical dilemmas. Some of these dilemmas are relatively easy to resolve, and some are not. Common ethical dilemmas that could occur include:

- Hazard Identification and Risk Analysis: Intentionally refraining from needed or relevant recommendations because of the fear of management response to them. Also, reducing the frequency or consequences of a scenario to avoid a costly or difficult action item.
- Auditing: Deleting or altering findings because of business concerns or embarrassment to those responsible for the PSMS. Also bowing to pressure from superiors to downplay the risk of audit gaps.
- MOC: Creating MOC records after a change has been implemented, or marking MOC action items as completed when they are not.
- MI: Definition of what is overdue for ITPM tasks and deficiencies so that the number of overdue ITPM tasks and open deficiencies is artificially low. For example, defining overdue as any tasks that were due by a certain date but ignoring those still not performed from before that date.
- Metrics: Definition of PSMS Key Performance Indicators (KPIs) to make the program appear to be in better shape than it really is. For example, liberal extension policies for overdue PHAs, incident investigations, audits, and other PSM-related action items so that they can be easily deferred so they are not captured in metrics.

4.4 EXTERNAL INFLUENCES ON CULTURE

Nearly all facilities interact with many external parties. Each may have a different culture than the facility, and exert an influence on the culture of the facility. Understanding these external cultures is important to encourage supportive external cultures and defend against cultures that could have a negative influence.

Some of the more common external parties that can influence culture include:

- Contractors,
- Labor unions,
- Vendors/suppliers,
- Industrial and residential neighbors,
- External emergency responders,
- Law enforcement,
- Regulators and elected officials,
- Trade and professional organizations,
- The media: and
- Financial institutions.

While not strictly external, corporate staff who work outside the facility and the company Board of Directors may also have cultures that differ from the facility.

The following pages discuss the potential influences of these external parties and how facility managers can manage those influences. Some of this material has been provided courtesy of Hoffman (ref 4.10).

Contractors

Contractors perform a wide variety of services for facilities, including operators, maintenance, construction, and professional services. When contractors arrive at a facility, they bring with them the culture of their company as well as cultural influences from other facilities they serve. Leaders should be aware of the degree the contractors' cultures differ from that of the facility. With that knowledge, leaders should then manage the business relationship to align the way contractors work and act with the facility's culture.

The economic forces that drive facilities to use contractors, and the cost- and time-competitive nature of contractors' business models are a fertile environment for *normalization of deviance*. This should be addressed with very specific contract language clearly stating the imperative for process safety and requiring procedures to be exactly followed. The contract and contractors them should be closely managed to those requirements (see discussion of Contractor Management in Section 5.1).

Ensuring open and frank communications can be a challenge with contractors. Labor law in many countries discourages companies from using contractors on a permanent or quasi-permanent basis. Such laws may inadvertently encourage companies to omit communicating with contractors or communicate via a roundabout route that makes it difficult to timely report safety issues. However, no law should be interpreted to limit communication with contractors about safety requirements and safety issues. Leaders should therefore include contractors in their safety communication chains.

Less experienced workers can be unduly influenced by more experienced contractors and vice versa. In either situation, the company's safety and health practices must be clearly understood and enforced. Also, the details of applicable codes and standards must be followed and any deviance from the contract or specification needs a review and approval. If an apparent conflict occurs, it should be raised to leaders for guidance.

Labor Unions

Labor unions were created to advocate for workers' rights and safety. Therefore, the labor-management relationship can serve as a platform on which to build the basis for a strong process safety culture. Furthermore, the process safety culture core principles directly address the relationship between management and workers for process safety culture and therefore can serve as a roadmap. Of the core principles, management and labor should find a mutual understanding of the *imperative for process safety. Mutual trust* and *open and frank communication* should also be established. From that basis, the labor contract can define how *individuals will be empowered to successfully fulfil their process safety responsibilities.*

If the management-labor relationship is initially strained, it could take time and effort to work through this process, especially to establish *mutual trust*. Ways to do this are discussed in section 2.3.

Leaders should keep in mind several common pitfalls that can exist in management-labor relations. These include:

- **Overtime vs. Fatigue:** There is a strong mutual incentive for overtime. Workers seek to increase their income and management seeks to limit the number of workers on the payroll. This can lead both to *normalize deviance* with excessive working hours that cause fatigue. Labor agreements should establish clear fatigue-based limits for overtime, to which management and labor should then adhere.
- **Workplace Involvement:** As discussed in section 5.1, workforce involvement is important to bring the front-line perspective to hazard and risk assessment, operating procedures, and other PSMS elements. Making this happen can be a challenge however. If workers are pulled from their shifts, process safety responsibilities may be left uncovered. Therefore, leadership should provide competent fill-in coverage, or involve the worker during over-time hours, subject to fatigue limits mentioned above.
- **Seniority:** Many labor agreements grant preferential treatment to workers who have been employed longer by the company. Commonly, longer-term workers may be

given first choice of work assignments and shifts. This, however, does not guarantee that the worker selecting the assignment has the necessary competence (see section 5.4). If a worker can choose an assignment without having the necessary competence, the probability of a process safety incident increases. While *empowering others to fulfill their process safety responsibilities* is desired, an employee cannot be empowered if they are not competent to carry out their role.

- **Labor actions:** During contract negotiations or disputes, labor unions may organize walk-outs, slow-downs, and other forms of protest. Any of these have the potential to impact process safety. In a strong process safety culture, labor should seek other ways of addressing concerns, leveraging *open and frank communications* as a starting point. Indeed, improved labor relations can be an additional benefit of efforts to improve process safety culture.

Vendors/Suppliers

Like contractors, the relationship of vendors and suppliers with a facility is fertile ground for *normalization of deviance*. Sometimes this can occur through the best of intentions such as in the following case study.

A vendor received an order for 20 carbon steel flanges. They filled the order for 19 of the flanges with the carbon steel as specified, but supplied the 20th flange from a "superior material," in this case 2.25Cr. When challenged by their customer for this discrepancy, they indignantly replied that they had supplied the superior material at no extra cost. The facility applied the wrong welding procedure for the 2.25 Cr flange, leading to a failure during operation (Ref 4.11).

The solution is also similar as for contractors. Purchase requisitions should state very clearly the specific process safety requirements for the goods purchased, highlighting the *imperative for process safety* and requiring the specification to be exactly followed. The purchased goods should be verified vs. this specification through positive material identification and other relevant measures. (see section 5.2, Asset Integrity).

The Public

A company with a strong process safety culture should consider the public's safety concerns. The public includes the residential and industrial neighbors of a facility but goes well beyond that. The increasing use of both traditional and social media has the potential to extend the public interested in any given facility considerably, including beyond the local area.

The term "license to operate" is sometimes used to describe the potential influence of the public on a facility. If the public is displeased with a facility for a wide range of reasons, not all process safety related, they can bring pressure on the facility via the media, courts, regulators, and local officials. This can lead to restrictions on expansions or modifications, reduction of operations, and other adverse business effects.

The PSMS element stakeholder outreach (see section 5.1) addresses the relationship the facility should develop with the public. From a culture perspective, facilities should develop an understanding of the core values of the public. While different groups may prioritize these differently, the public's core values tend to follow this descending order of importance (Refs 4.12, 4.10):

- *Health and safety of themselves and their families.* The public tends to care most about their health and safety and the health and safety of their families, particularly their children.

- *Value of their property and possessions.* The public worries about incidents and environmental impacts that could damage their property. They are also concerned about the potential influence of the facility on their property values.
- *Environmental protection.* Most people regard themselves as pro-environment in some way. This may take many forms, from simply appreciating nature to actively protesting. They are concerned about "What YOUR plant is doing to OUR environment."
- *Quality-of-life.* This core value encompasses three objectives:
  1. Pride in Community, including the aesthetics of their neighborhood and nearby businesses,
  2. Absence of Conflict. People do not enjoy fighting over issues such as chemical releases (including nuisance odors), frequent truck traffic, or rail crossing delays, and
  3. Freedom from Fear, the absence of constant concern about what events might occur in the middle of the night or while their children are in school.
- *Economic security.* Beyond the value of property and possessions, the public is concerned with how the facility affects the overall economic condition of the community. This can include employment of family members and friends, contributions to local community organizations, and producing products or raw materials needed by other local businesses. Anything that happened to the facility could potentially impact community economic security.
- *Peer pressure conflicts.* If a friend or neighbor feels that the facility threatens a value important to them, they may start some type of community action, such as a petition against the facility. This can then lead to additional discomfort in

the community as the petition is circulated. While some may sign willingly, others may weigh the benefits against the concerns differently. This can create discomfort and friction between members of the community. It can also undermine confidence in agreements between management and community leaders and degrade the trust that has been developed over time.

- *Moral and ethical principles.* The public sometimes has a low tolerance for behavior that they find reprehensible for moral or ethical reasons (Ref 4.10). This behavior may be legal or illegal, and may or may not directly impact community members. However, if the conduct impacts a sizeable number of people or has a serious impact, the public can lose confidence in institutions in general.

These public core values inform the culture of the public with respect to any business, particularly those with facilities that pose significant public or environmental risks that are near their community. The public values tend to be consistent with process safety culture core values. However, the way the local public values exist must be understood and appreciated by a facility. This helps the facility translate the way it talks about its values into terms that resonate with the way the public talks about its values. The facility should engage in several ways with the culture of the public to remain aware of shifts in public sentiment and not be surprised by big changes. This is a lesson learned the hard way by the nuclear power industry following the Three Mile Island accident in 1979 and by the process industries following Bhopal in 1984.

The process to stay connected with the public and harmonize public values with company culture starts internally. First, organizations should develop a culture that recognizes their unofficial license to operate is real and must be earned. Organizations also need to get beyond the feeling that they are being treated unfairly when public or media attention of them is adverse. Once these biases are overcome the organization can proceed to interface with the public and be a positive impact on their culture. If a facility has been a good corporate citizen, they have a much better chance of being able to weather the challenging times when an incident occurs. A good corporate citizen can affect the culture of the public in the following ways:

- Supporting the local community in tangible ways. This includes both financial donations and volunteering.
- Supporting the local schools by providing information about which they are expert and the time of their employees for presentations, tours, and other information outreach.
- *Being open with information about their risks and* hazards and not being afraid to dialogue about them.
- Above all, being honest, credible, and consistent.

External Emergency Responders

Most facilities rely at least in part on outside emergency responders. These may come from nearby facilities as part of a mutual aid arrangement between plants or from public fire and ambulance services. All emergency responders, regardless of their origin, are motivated to respond quickly, hoping to limit damage, and to address the emergency aggressively. This stems from the public's expectation of emergency responders to protect them, and is codified in the legal principle of Duty to Act, which many see as requiring such aggressive response to emergencies. This principle may or may not be written into the national or local law, but its influence exists regardless. Emergency responders should research how their national and local laws define a responder's duty to act both on- and off-duty. (Ref 4.13)

Experience has shown however that aggressive response may, in many cases, be the worst option. This was reinforced by the retail fertilizer facility explosion in West, Texas, USA. Twelve firefighters aggressively fighting a building fire were killed when stored ammonium nitrate detonated (Ref 4.14). Emergency response experts agree that the correct action would have been to stand back and evacuate the nearby public, as was done with the St. Louis, MO, USA cylinder facility fire (Ref 4.15)

To incorporate external emergency responders into the facility process safety culture, it is important to align them with culture principles of *understand and act on hazards and risks*, and help them *maintain a sense of vulnerability*. Some external emergency responders may come equipped with that culture. This is more likely with large municipal fire companies and in industrial fire companies involved via mutual aid agreements. In such situations, companies tend to drill with the external responders and provide the necessary information in advance about hazards, locations, and procedures.

For the external responders who do not already have this culture, the facility should work in advance to establish it. This group includes large municipal fire companies and mutual aid companies that are not already acculturated as well as all others, with added emphasis on volunteer fire companies.

The desired attitude of emergency responders should be, "If You Don't Know, Don't Go." To help responders know and help them make correct decisions, companies should provide Safety Data Sheets (SDS) and plot plans for relevant materials in advance, and should placard trucks and label tanks so they can be readily identified. Even with this information, facility leaders should coordinate with external responders before they enter the facility to discuss the appropriate type of response.

To integrate the culture of offsite emergency responders with the facility culture, leaders should start by educating them about the facility and its culture. This can be accomplished in several ways:

- Invite them to the facility for briefings, tours, and indoctrination about what the facility does, the hazardous materials handled and where they are located, and the emergency response plans.
- Share information about personnel, training, equipment, etc. and discuss gaps that need to be addressed by the responders and by the facility. This is particularly important when the facility relies mainly on external responders
- Discuss how response duties will be shared. Facilities that depend on external responders will still need to identify points-of-contact within the facility to advise on hazards, locations, headcount, etc.
- Hold joint emergency exercises, both table-top drills and mock scenarios within the facility. Sometimes the budgets or other commitments of outside responders will limit the extent of such exercises, but the invitation should be regularly extended nonetheless.
- Coordinate and exercise communications channels regularly between the onsite emergency response team and offsite responder organizations.

It is important that the interface with local responders, in whatever form it takes, should be accomplished in advance. If the first time a group of offsite responders have seen the facility is when they arrive at the front gate in response to a live event, the relationship will not start off well.

Law Enforcement.

The role of law enforcement has traditionally been related to preventing and responding to criminal activity, which rarely interfaces with matters of process safety management systems or culture. Law enforcement also becomes involved in emergency response, especially if community evacuation is required. When responding to incidents, the role of law enforcement mirrors that of emergency responders. For this purpose, law enforcement should be included in relevant discussions with emergency responders and emergency planning exercises.

Since September 11, 2001, the process industries have come to realize that terrorist attacks can have similar consequences to process safety incidents. The role of law enforcement in preventing terrorist attacks creates the need for another kind of relationship with the facility. Facility security requires different activities than process safety, but shares the need for coordination with and outreach to law enforcement.

Regulators

At first glance, it might seem that regulators have no role in the process safety culture of the facility. An inspector arrives at the facility and is greeted respectfully, but guardedly. The inspector may or may not find compliance gaps, then the findings may or may not be challenged by corporate counsel. A decision is reached, the facility takes the required measures, and that is the end of it. Building a good relationship with the regulators, including good one-on-one relationships with the inspectors could pave the way to amicably resolving disagreements about compliance findings. However, this is a one-on-one relationship between company compliance professionals and the regulators that tends to exist outside the culture.

The one-on-one relationship can be taken to the facility level and begin to influence the facility culture. In many countries, the Responsible Care® program of the International Chemistry Council, operated in many countries by the national chemical industry trade association, encourages facilities to build partnerships with regulators. Responsible Care leverages a proactive approach to process safety as well as other environment, health, safety, security, and quality, and encourages collaboration with regulators. This can help strengthen PSMSs and contribute towards *establishing the imperative for process safety.* Various country and regional programs led by regulators such as OSHA VPP (USA), Safer Together (Australia), and Step Change in Safety (UK) seek similar goals.

The threat of a routine regulatory inspection is generally not an incentive to improve culture or PSMS performance. In general, regulatory agency staffing levels are rarely sufficient to put teeth in such a threat. Some agencies have been trying to change this by focusing on only 1-2 PSMS elements and certain sub-sectors, such as the National Emphasis Programs (NEPs) used in recent years by OSHA in the USA. Facilities leaders should take care to prevent regulatory focus on just a few elements from leading to *normalization of deviance* or loss of the *imperative for process safety* in the other elements.

While some process safety regulators around the world are themselves process safety experts, the majority are not. Their backgrounds may be in occupational safety, environmental sciences, or similar disciplines that enable them to interpret regulations and understand management systems. In other words, regulators will generally not conduct in-depth technical analysis, but they will understand and evaluate management system performance. They will also be sensitive to cultures that do not take management systems performance seriously.

Regulators will certainly come to the plant following a major release incident. In such cases, regulators will generally have one or more regulatory findings. Having a collaborative relationship with the regulator while demonstrating a strong culture will help limit findings by keeping the regulators' focus on the relevant portions of the relevant PSMS elements, and help prevent the findings from being elevated to willful or serious violations.

In summary, regulators will be part of the process safety culture, whether encouraged to do so or not. In a strong culture, leaders should engage collaboratively with regulators to build trust, and then demonstrate the strong culture (and PSMS performance) for the best regulatory outcomes.

Elected Officials

Elected officials will tend to provide positive reinforcement to a facility that is widely viewed as a valued member of the community that elected them. As discussed above, providing jobs, having a low impact on noise, odor, and the environment, and avoiding process safety incidents all contribute to a facility's positive public image. Having a collaborative relationship with elected officials can help reinforce that positive image. The positive external reinforcement coming from elected officials can be a source of pride for employees that helps sustain a positive culture.

Conversely, when the public image of the facility turns negative, elected officials can quickly change their feelings for the company and express those feelings widely through the media. The resulting negative press can depress an already weak culture. Stronger cultures can weather the criticism of elected officials if leadership shows *trust* that employees know the *imperative for process safety* and expresses confidence that the situation that created the negative image will be addressed.

The Media

The media is for the most part a vehicle for elected officials, regulators, and the public to express their opinion about the facility, whether positive or negative. To that extent, the comments above regarding these groups apply to the media as well.

When incidents happen, however, the media has the potential to have impacts that go beyond their role as a communication vehicle. Media may sensationalize bad news out of traditional journalistic practices or to promote readership or viewership. Media frequently over-simplifies stories, omitting facts that that could have placed the facility in a better light. They may also repeatedly run stark images of the worst of the incident with dramatic voice-over, reinforcing a negative image in the minds of the reader or viewer (Ref 4.10).

They may do this for innocent reasons such as to meet a deadline, to fit space available, to make the story broadly understandable, or because they are waiting for more information. However, in recent years, numerous publishing, broadcast, and social media outlets have emerged that intentionally slant the news. Some of the slanted news could be more favorable to the facility, while others could be less favorable. And today, it is not unusual for anyone with a cellphone to capture video and broadcast their view of what they are seeing.

Generally, the facility should engage with the media and not try to avoid it or stonewall. Trying to fight or avoid the media will typically backfire, resulting in the negative impact of the simple phrase "The facility refused to comment."

Facility leaders should be aware that media may attempt to contact employees, hoping to get an inside story about an incident or a sound bite that supports the story. It is impractical and not even desirable to provide all employees with media training. However, if the process safety culture is strong, the employees' commitment to process safety should come through.

Trade and Professional Organizations

Trade and professional organizations fulfill several roles, but generally support the company's interactions with the public, the regulators, elected officials, and the media. Trade organizations generally focus on improving business conditions through activities such as lobbying and general marketing and publicity for the industry. Professional organizations focus on workers in the profession, providing training, guidance, standards, certification, and other services that help them perform their jobs.

Generally, membership in trade organizations is on a corporate or facility basis, while membership in professional organizations is on an individual basis. However, many exceptions exist, and some organizations have characteristics of both kinds of organizations. For example, some trade organizations with corporate membership also develop standards as a professional organization would, using committees with individual membership. Likewise, some professional organizations have corporate membership options, and some seek to raise the image of the profession and by extension the industry.

Companies having or seeking to develop a strong process safety culture should be actively involved in a variety of trade and professional organizations. In addition to helping interactions with the public, regulators, elected officials, and the media, participation exposes them to emerging trends, opportunities to improve, lessons learned and even negative examples to avoid. Participation should be considered at the global, national, and local level, to involve a variety of leaders and obtain different perspectives.

Financial Institutions

Financial institutions include banks and other lending organizations, insurers, holding companies, bond and stock holders. Most of these institutions operate on objective criteria such as actuarial principles, accounting rules, and profitability expectations. These criteria may be perceived to conflict with the *imperative for process safety*, and may inadvertently encourage *normalization of deviance* as discussed previously.

However, PSMS performance and culture have a key role in satisfying the demands of financial institutions. All are interested in managing risks of all kinds, including process safety, as risk represents a threat to expected financial performance. Higher actual and perceived risks can also drive other financial factors such as interest rates and insurance premiums. In other words, financial institutions may not care about the details of the PSMS or how leaders build culture, but they are keenly interested in the capability of the PSMS and culture to manage risk.

These institutions are also interested in the reputation of the company and the confidence in management's ability to maintain performance and grow. Positive performance in these areas can help attract investors and increase the stock price. A serious incident or a series of smaller incidents can erode that confidence and drive down shareholder value.

In summary, leaders should beware of misinterpreting the pressure from financial institutions. They should resist the natural inclination to seek shortcuts to increase short-term profits and instead recognize that by driving for strong process safety culture and PSMS performance, they will ultimately provide the performance these institutions seek.

<u>Corporate Staff</u>

Ultimately, companies should strive for consistent process safety culture throughout the organization. However, a company's facilities and indeed the corporate process safety team may be in different places on their culture improvement journey.

The corporate process safety team and the corporate leaders they support can sometimes have a different view of culture, the PSMS, and the underlying technology than their facility counterparts. Some corporate staffs exert close, centralized control over their facilities' PSMSs, providing detailed policies and procedures and allowing little flexibility. Others take a decentralized view where the corporate team provides high-level expectations and governance, and leaves implementation to the facilities. Most companies operate somewhere between these extremes.

Companies on the centralized end of the spectrum will tend to drive culture improvement from the top down. This can help establish the *imperative for process safety* from the top down, but may detract somewhat from the sense of *empowerment (to successfully perform process safety responsibilities)*. Corporate teams should recognize that not all facilities will be at the same place in every core cultural principles. Some facilities will be ahead on some principles and other facilities on other principles. Likewise, none may be at the same place as the corporate group. The corporate leadership role should recognize that each facility will need a different kind of support, and perhaps encourage facilities with complementary strengths to help each other.

Decentralized companies may find that some facilities and indeed some businesses are superstars in culture improvement, while others have trouble getting started. Often this can be addressed through business leadership. However, it may require the corporate group to step in and provide more support and one-on-one leadership

Board of Directors

The responsibility of boards of directors is to the company's shareholders, not to company management. Most shareholders generally have a single concern to the exclusion of all others –

steadily increasing share price. Therefore, boards are heavily influenced to focus their attention on matters that relate directly to the share price. As discussed relative to the financial community, this can unintentionally motivate management to *normalize deviance.*

Some boards are more independent or more supervisory than others. Companies with weaker process safety cultures often have boards that acquiesce in management decisions and take little interest in process safety. Companies with stronger process safety cultures have boards that recognizes that a strong PSMS and culture will help reduce risk and protect the company image, two things that help increase share price. Boards should oversee auditing programs to ensure that they have a true measure of the health of the company's process safety effort. Towards this, some companies have found it helpful to have one or more board members who understand process safety concepts. This approach is expected to gain in practice in the coming years.

4.5 Process Safety Culture Metrics

Like other aspects of a PSMS, process safety culture should be measured periodically to monitor progress, guide improvement, and detect regression. Culture changes, especially positive changes are usually slow and hard to discern. Nonetheless, historic experience demonstrates that regular monitoring can reveal cultural changes over time.

Practical experience as well as formal study by the International Atomic Energy Agency (IAEA) (Ref 4.16) has shown that a single quantitative measurement of culture may be impossible. Instead, culture can be sensed from qualitative indicators. Therefore, facilities and companies should select a range of indicators that reflect the individual culture core principles. These indicators may be based on observable behavior, conscious attitudes, perceptions, or beliefs

The following potential culture metrics have been culled from Guidelines for Risk Based Process Safety (Ref 4.17), the work of the IAEA mentioned above, and the experience of subcommittee members. Many of the potential metrics discussed below can be collected from operating and other business records. Others rely more on anecdotal evidence collected from interviews and observations. Companies and facilities should select a modest number of indicators that best fit the core principles that need the most improvement or that, from experience, represent the most sensitive areas. Most companies should consider including metrics related to the *imperative for process safety* and *preventing the normalization of deviance.*

Establish an Imperative for Safety

- Has the independence of the PSMS changed and if so, how? Independence can be measured by the presence of conflicts of interest. For example, an organization change resulting in the process safety manager or chief inspector reporting to the Operations Manager would represent such a conflict. These formal indicators should be easily detected. More informal losses of independence can occur if a long-term, well-respected process safety manager whose opinions on process safety were respected by management retires and is replaced by a much younger unknown hired from the outside. Although the new manager may be knowledgeable and even eager to perform in this role, there may be a loss of influence until the new person builds the same reputation as their predecessor.
- The increase or decrease of total resources devoted to the PSMS in a fixed period. Resources in this context refer to budgets, numbers of persons with direct PSMS responsibilities, and contractors who perform important process safety duties. Because resource loads can vary

over the year, this indicator should be measured no more than semi-annually or annually.

- Are process safety metrics reviewed by management with the same attention that is paid to production or quality related metrics?
- How many times in the time period have employees with key process safety roles been placed in the undesired position of having to prove that an operation, design, potential hazard scenario, or some aspect of the operations is unsafe? Avoiding this situation is one key outcome sought through the imperative for process safety. Collecting this metric may take some effort, as there could be many people potentially exposed to such pressure.
- The number or percentage of startups or re-starts after an abnormal situation approved with pending safety issues, such as overdue ITPM, bypassed safety features, etc. This is a measure of how many times production has taken precedence over safety. If zero because these issues were not tracked would also be a negative indicator.

Provide Strong Leadership

- Frequency or number of senior manager visits the worksite, or percentage of the scheduled visits that actually take place. A low value may indicate that upper management places little value on personally motivating strong process safety performance.
- Percentage of managers and supervisors who have been trained on creating and maintaining a strong process safety culture. A high value is indicative of the value that the organization places on implementing positive process safety culture change.
- Percentage of meetings addressing process safety topics that include active participation by a member of upper management. Alternatively, percentage of leadership team

meetings including substantial discussion of process safety. An imbalance in emphasis may be indicative of a management attitude that process safety is less important.

- Frequency with which relevant process safety statistics are shared with the organization. A low value may indicate that management does not adequately appreciate the value of informing the workforce of the organization's process safety performance.
- Manager attendance at management review meetings. Poor attendance may indicate a low interest in process safety performance, or in communicating management expectations.

Foster Mutual Trust

- This core principle can be subjective, and leaders and workers may have a different opinion regarding trust. Those opinions may be difficult to elicit from fixed surveys. Therefore, this core principle should be assessed primarily by interviewing leaders and workers. Generally, the interviews should consider whether interviewees feel that:
  o A just system exists where honest errors can be reported without fear of reprisals,
  o Submitted information will be acted upon,
  o Bad ideas can be challenged, discussed, and resolved satisfactorily; and
  o Errors will not be punished unless the act was reckless, deliberate, or unjustifiable.
- Since trust between peers is also important, the same approach can be applied to peer interactions.

Ensure Open and Frank Communications

- Do employees exercise stop-work authority? When they do, does leadership thank them and take care to avoid second-guessing their decisions?

- Is shift turnover being conducted formally as prescribed in facility or company procedures? This can be confirmed by observing the activity, taking care so that the operators involved are not role-playing for the observer. If the procedures require log entries documenting the turnover, then these can be checked and the percentage the log entries were documented correctly can be calculated and trended.
- Regular scheduling and attendance of health and safety committee meetings by facility leadership. Agendas and attendance records of these meetings should allow this metric to be trended.

Maintain a Sense of Vulnerability

- Number of process safety near misses reported and recorded during a fixed period. Increases or decreases in this number can be either bad or good so it is also important to understand the reason for the trend. For example, a decrease could arise from complacency in reporting, or could represent fewer actual near misses, and conversely for an increase.
- Does the organization challenge its metrics? Good metrics could indicate actual improvement, or could reflect complacency in reporting, so skepticism of good metrics is a positive indication. However, challenging or trying to justify poor metrics is an indication of culture weakness. This indicator is qualitative and somewhat subjective.
- Are the facility's vulnerabilities on everyone's minds? This can be sensed by asking employees what are their biggest process safety concerns. Usually employees will talk about the execution of one or more PSMS element. With a strong *sense of vulnerability*, the response will also note vulnerabilities. For example, "I'm concerned that we keep up on inspections, so we do not have a release."

- The percentage of process safety incident investigations during a fixed period where the management system and technology root causes are identified. In general, operator error would not be counted. Not finding root causes or finding operator error as a cause indicates weak commitment to process safety, complacency, and a poor sense of vulnerability.
- The quality of HIRAs/PHAs is also an indicator of the sense of vulnerability. Obtaining an independent assessment of completed HIRAs/PHAs can determine if methods are being kept up-to date and performed with sufficient thoroughness. The review would look for non-conservative assumptions, missed scenarios, and invalid safeguard credits.
- The reaction to weak signals is a good indicator of sense of vulnerability. Is there a low threshold for intervening, investigating, or stopping work if something does not seem right? How frequently does this happen? Or are "false alarms" commonly accepted. This is another qualitative measure that can serve as an important benchmark for organization complacency.
- Have organization personnel begun to lose respect for process hazards/risks? This would mainly be detected by observation and anecdotal evidence. Some observable indicators include PPE violations, caps left off low point drain valves, external evidence of corrosion, drips, residues, and violation of access controls. Also note differences in behaviors between employees and contractors.
- Is the attitude of personnel shifting to "can-do," "get it done fast," or heroics? This is often accompanied by frequent by-passing of critical safety systems with or without supporting analysis, heroic measures counter to emergency procedures (e.g. fighting fires, attempting to

shut down, or attempting rescue, instead of evacuating). Many of these behaviors can be tracked quantitatively.

Understand and Act Upon Hazards/Risks

- The number of HIRAs/PHAs that are overdue and the aging of these overdue studies.
- The number of action items, such as recommendations from PHA/HIRA, incident investigations, audits, etc. that are overdue and their aging.
- Evidence that hazards and risks are disregarded in PHA/HIRA, MOC, and operational readiness. This can take the following forms:
  o An attitude of "it cannot happen here,"
  o A common belief that two safeguards cannot fail simultaneously,
  o A pattern of over-stating the risk reduction of a safeguard
  o A pattern of multiple non-independent safeguards; and
  o An over-reliance on human intervention as a safeguard.
- The number of times a risk level deemed unacceptable by HIRA/PHA or MOC have been de-facto accepted by failing to implement recommended safeguards? Similarly, the ageing of action items related to risks determined to be in the category of "reduce as soon as possible." Both will indicate whether management is taking the risk criteria seriously. If PHAs are performed correctly, this may be determinable by tracking closure of PHA recommendations. If the facility is not sure of PHA quality, the previous metric should also be considered.
- Belief that the absence of regulatory citations, accreditation or recognition by ISO or similar organizations means the PSMS is adequate.

- Does the scope and boundaries of the PSMS cover all hazardous materials or processes or are some omitted because they are not covered by regulation?
- Is the facility using recognized good engineering practices to control hazards and maintain equipment and safeguards? In the USA. and other countries that follow the PSM regulation, are RAGAGEPs identified and followed?
- Are people with the right experience, skills, and perspectives being assigned to HIRA/PHA, MOC review and incident investigation teams? Gaps indicate weakness in ability to understand and act upon hazards and risks.
- Is process safety knowledge up-to-date? Measurements can include number of completed MOCs where process safety knowledge has not yet been updated, the length of time following MOC to update process safety knowledge, and the length of time since the last update.
- In more advanced cultures, have the risks determined through HIRA/PHAs been used properly to determine the levels of effort and evaluation of other RBPS elements? A key aspect of RBPS is that processes and operations with higher risk should receive greater attention in the asset integrity effort, have a more detailed and higher level of approval in MOC and Operational Readiness, more specific training, etc. Likewise, lower risk processes and operations may receive less attention, but of course should not be ignored.
- Evidence that the permit-to-work process is insufficient. A review of completed permits should show whether job safety analyses were performed with adequate risk analysis and appropriate isolation and safeguarding were performed.
- Evidence of inadequate MOC. This can be determined by comparing change orders to MOC documents to look for changes claimed to be replacements-in-kind that were not

and other changes that should have been subject to MOC where MOC was not done.

- Extent to which emergency plans contain inadequate actions. This may include designations to shelter-in-place where the shelter is inadequate or not accessible, evacuation routes that are unsafe, etc. This can be examined critical through the audit process.

Empower Individuals to Successfully Fulfill their Safety Responsibilities

- How many times has an operator initiated a shutdown of process equipment on their own when warranted? This indicator can be determined from operating logs, incident reports, and possibly the distributed control system, as well as interviews. In evaluating the trend over time, leaders should be careful to understand the reasons behind the trends. An increasing trend could signal greater empowerment, but also could signal that the process is getting less stable and needs to be shut down more frequently.
- If the organization has a stop work authority policy, how many times has it been exercised over a fixed amount of time? If this authority is formally granted in writing, there should be records available that show when it has been used, i.e., operating logs, incident reports. Even in the absence of a formally granted Stop Work Authority, interviews and documentation reviews should reveal if facility personnel took it upon themselves to halt an operation, maintenance task, or other activity that they believed presented clear and present danger.
- How many required training sessions are overdue and what is the aging of these overdue requirements? The trend of this data can also be reviewed over time to determine if the situation is improving or worsening.

- How many training sessions and opportunities have been cancelled or rejected by management over a fixed period and has that number been increasing or decreasing? This will be an indication of how training resources are being allocated.
- Average response time to the resolution of a process safety suggestion. Slow management responses to suggestions may provide a disincentive to employee participation.
- Number of process safety suggestions reported each month. A low value may reflect a low level of employee engagement in improving process safety, or a perception that employee participation offers a low return on the investment in effort.

Defer to Expertise

- The emergency shutdown and stop work authority metrics mentioned under Empower Individuals to Successfully Fulfill their Safety Responsibilities above apply here as well.
- Other aspects of this core principle can be subjective, and leaders and experts may have a different opinion on the extent that expertise is deferred to. Those opinions may be difficult to elicit from fixed surveys. Therefore, this core principle should be assessed primarily by interviewing leaders and experts. Generally, the interviews should consider:
  - o Comparing the degree to which experts feel they are deferred to by leadership and can influence decisions to the leaders' perception about deferring to the experts. Areas where perceptions are different represent improvement opportunities.
  - o Whether differences in perception between leaders and experts exist depend on the area of expertise or

the PSMS element. This may identify specific leader-expert interactions that should be improved.

o   Has a Technical Authority system been established and is it working? Are process safety and technical experts engaged?

Combat the Normalization of Deviance

*   Typically, this can be seen from PSMS element metrics. The trend in the element metrics will indicate whether normalization of deviance is decreasing or increasing. For example, monthly or yearly trends in:
    o   Overdue asset integrity tasks.
    o   Open asset integrity deficiencies.
    o   Overdue process safety action items.
    o   Number of bypassed or removed safety features that have exceeded their expiration date.
    o   Overdue process safety training activities.
    o   Changes that should have been considered by the MOC process but were not.
    o   Incident investigations that show failure to follow procedures as a contributing factor or root cause.
    o   Nuisance alarms. Ignoring these can lead operators to ignore real alarms.
*   A subtler, yet important indicator of the normalization of deviance is when operators disbelieve instrument indications. This can happen because the instrumentation is chronically out of calibration or inaccurate, misleading. In such cases, operators may be hesitant to take timely actions including emergency shutdown. This information may be found in DCS logs but may be more quickly revealed through interviews.
*   The number of process safety near misses that have been reported during a fixed period. Has that number been going up or down? This metric is widely valuable as

discussed throughout this book, but it is particularly useful in identifying normalization of deviance.

- Fatigue, resulting from excessive overtime, can lead to conditions conducive to normalization of deviance. Overtime records can be trended in various ways, including cumulative overtime, number of extended shifts, and fraction of workers extending their shifts in each time period.

<u>*Learn to Assess and Advance the Culture*</u>

- In many ways, the actions taken resulting from metrics for the other culture core principles indicate how well the facility and company is learning, assessing, and advancing the culture.
- Participation in voluntary process safety activities within the company and in trade and professional groups indicates the degree to which learning from outside the company is being considered.

4.6 SUMMARY

The process safety culture of the organization depends heavily on human behavior. Leadership can influence this behavior positively or negatively, as can many outside influences. Ethics can be a motivating force, especially if ethical behavior is modeled by leaders. Compensation can play a role in driving the desire culture, however, leaders should exercise care to prevent compensation from unintentionally driving undesired behavior.

Assessing the existing culture and then implementing changes to correct it can be challenging, but ultimately should be done to focus efforts where they can make the biggest difference. Ultimately, the application of the core principles of process safety culture is a journey. Leaders and employees need to put in the work to build a strong culture. There are no shortcuts.

4.7 REFERENCES

4.1   Center for Chemical Process Safety (CCPS), *Guidelines for Preventing Human Error in Process Safety*, American Institute of Chemical Engineers, 1996.

4.2   Daniellou, F., Simard, M., Boissières, I., *Human and Organizational Factors of Safety - State of the Art*, Fondation Pour une Culture de Securité Industrielle (Foundation for an Industrial Safety Culture), 2011.

4.3   Choudhry, R., et al, Safety Science 45, Elsevier, 2007.

4.4   Krause, T., *The Ethics of Safety*, EHS Today (http://ehstoday.com/safety/best-practices/ehs_imp_67392), June 2007.

4.5   Morris, W., et. al., *The American Heritage Dictionary of the English Language, New College Edition*, Houghton Mifflin Co, 1976.

4.6   McNamara, C. Complete Guide to Ethics Management: An Ethics Toolkit for Managers, http://managementhelp.org/businessethics/ethics-guide.htm,

4.7   Grubbe, D., *Ethics – Examining Your Engineering Responsibilities*, Chemical Engineering Progress, Vol. 111, No.2, American Institute of Chemical Engineers, February 2015

4.8   McDonald, A., *Truth, Lies, and O-Rings, Inside the Space Shuttle Challenger Disaster*, University of Florida Press, 2009

4.9   American Institute of Chemical Engineers, Code of Ethics, http://www.aiche.org/community/sites/local-sections/sts/code-ethics.

4.10 Hoffman, J., *Keeping Cool on the Hot Seat – Dealing Effectively with the Media in Times of Crisis, 5th Ed.*, 2011.

4.11 Kletz, T., *What Went Wrong? Case Histories of Process Plant Disasters and How They Could Have Been Avoided*, 5th Ed., 2009.

4.12 Lucaszewski, J., *Influencing Public Attitudes: Strategies that Reduce the Media's Power*, 1992.

4.13 Mangeri, A., *Duty to Act: Legal Obligations vs. Community Expectations*, American Military University, 2014.

4.14 Chemical Safety and Hazard Investigation Board, *Investigation Report – West Fertilizer Company Fire and Explosion*, 2013

4.15 Chemical Safety and Hazard Investigation Board, *Safety Bulletin – Dangers of Propylene Cylinders in High Temperatures*

4.16 International Atomic Energy Agency, *NS Tutorial, Section 6., Developing Safety (6.2.1 How to Measure Safety Culture)*, 2001.

4.17 Center for Chemical Process Safety (CCPS), *Guidelines for Risk Based Process Safety,* American Institute of Chemical Engineers, 2007.

5
ALIGNING CULTURE WITH PSMS ELEMENTS

Previous chapters have noted that process safety culture is an essential factor in the success of a company's or facility's Process Safety Management Systems (PSMS). In fact, culture influences each element in a PSMS and can make the difference whether that element succeeds or fails.

Companies use a wide range of PSMSs that may be designed based on regulations, trade group practices such as Responsible Care®, quality standards such as ISO-9001, or company business management practices. Many companies follow the Risk Based Process Safety (RBPS) approach (Ref 5.1 CCPS), base their system on it, or developed something similar. Since RBPS spans nearly all considerations addressed by the PSMSs in use, this chapter will use it to discuss the alignment of process safety culture with the PSMS.

RBPS is based on 20 elements, organized into 4 foundation blocks that link it to the quality principle of Plan-Do-Check-Act. Figure 5.1 shows the organization of the 20 RBPS elements by foundation block.

As discussed throughout this book, culture starts with senior corporate leadership. It is then reinforced by leaders in all other levels and functions.

Essential Practices for Creating, Strengthening, and Sustaining Process Safety Culture, First Edition. CCPS. ©2018 AIChE. Published 2018 by John Wiley & Sons, Inc.

Figure 5.1 Risk Based Process Safety Management System

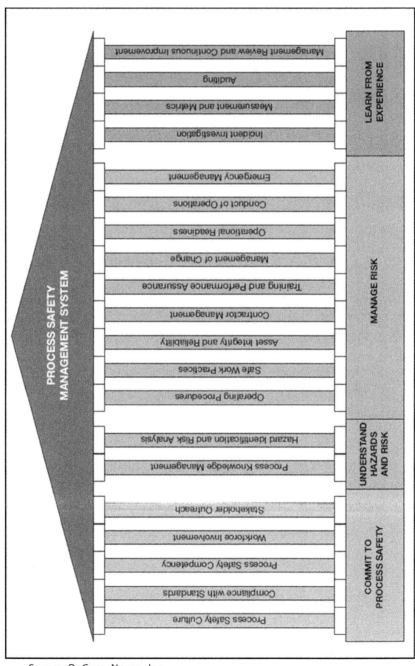

Source: D. Guss, Nexen, Inc.

In most publications, the RBPS elements are discussed in sequential order. However, to help senior leaders and other functional groups (e.g. process, risk managers, and workers) navigate their culture roles, this chapter presents the elements by the role where they may make the biggest culture impact. The senior leader group of elements will be presented first, followed by the other three groupings. Table 5.1 describes these groupings.

These groupings are not intended to imply that leaders have no role in the other elements, nor that other functions do not have

Table 5.1 CCPS RBPS Elements Grouped by Role

RBPS Element	Senior Leaders	Risk Mgmt.	Process	Workers
Leadership	◆			
Standards			●	
Competence				★
W.F. Involvement	◆			
Stakeholder Outreach	◆			
Proc. Knowledge Mgmt.			●	
HIRA		▲		
Operating Procedures				★
Safe Work Practices				★
Asset Integrity		▲		
Contractor Mgmt.	◆			
Training				★
MOC			●	
Operational Readiness			●	
Conduct of Operations	◆			
Emergency Mgmt.t		▲		
Incident Investigation		▲		
Metrics	◆			
Auditing		▲		
Management Review	◆			

roles in the leader element grouping. Readers are urged to review the groupings and regroup as appropriate to their culture.

5.1 SENIOR LEADER ELEMENT GROUPING

This grouping of elements represents the process safety elements that senior leaders interact with most closely. That is not to say that leaders do not have a role in or interact with the other elements. As mentioned elsewhere throughout this book, senior leaders are accountable for all aspects of the PSMS. For these elements leaders tend also to be directly responsible for them or for managing those directly responsible.

Process Safety Culture (Element 1)

Experience has shown that without an imperative for process safety, established by senior management, and backed up by felt leadership, the best PSMS is bound to fail. CCPS therefore has long recommended culture as the very first element in the PSMS. Since Chapter 3 discussed the Leader's role in process safety culture, that discussion will not be repeated here. Readers who have not yet read Chapter 3 are recommended to do so before proceeding with the rest of Chapter 5.

Workforce Involvement (Element 4)

"That's all well and good, but let me tell you what really happens Sunday at 3:00 AM…" This line has been recited by front line personnel across the industry to engineers and managers for as long as the industry has existed. This clearly demonstrates that the workers who interact closest with the process have a view of how the process and management systems that may be closer to reality than the engineers and managers who designed and manage it.

Workers frequently are in a position to recognize warning signs and improvement opportunities that others who work away from the front line cannot see. The goal of the PSMS is to systematically engage with workers on the front line to understand these warning signs and opportunities. Establishing a strong process safety culture will help this happen by *fostering mutual trust* and by *ensuring open and frank communication.*

When senior leaders visit the workplace, they should engage with workers about process safety, put them at ease, and encourage them to speak freely. Then, following the culture principle *Understand and Act on Hazards/Risks*, the input should be acted upon. Just as importantly, senior leaders should insist that the other leaders in their organization do the same. This could take the form of ad hoc discussions on the plant floor, or formal workforce involvement meetings.

Such interactions need not focus exclusively on process safety. The principle of workforce involvement can also help identify warning signs and improvement opportunities related to quality, productivity, occupational safety, etc. Workforce involvement also has the potential to improve labor relations, easing future negotiations.

Workforce Involvement should ensure that the employees closest to the process hazards realize protecting their safety and welfare is the primary goal of the PSMS and their input is not only desired, but is imperative to the PSMS being effective. Workforce Involvement and PS Culture work hand-in-glove to build cooperation and trust among all workgroups.

Organizations that fail to achieve workforce involvement stand to lose more than first-hand knowledge of warning signs and improvements. Without workforce involvement, prime opportunities to build trust and open communication channels are lost. This can lead workers to believe that process safety is someone else's job, undermining the *imperative for process safety*. And as seen in the Columbia case history (Section 2.4), workers may decide not to report an actual serious situation because they believe their report will be ignored.

Stakeholder Outreach (Element 5)

Stakeholders generally mean members of the community near the facility, including residents, businesses, and public safety, security, and health organizations. However, there may be many more stakeholders that facilities and companies may want to or should interact with. These include trade, technical, and civic associations, suppliers, customers, and the industry at large.

Community stakeholders may be affected if a catastrophic incident occurs, and therefore should be prepared to act appropriately. Neighbors should know whether to shelter-in-place or to evacuate, and how to do it. Emergency responders should be prepared to address the consequences of the incident, from first response to intervention to medical treatment.

Beyond the community, outreach to suppliers can help assure that no new hazards are introduced to the facility though changes to incoming materials. Outreach to customers can help them manage the hazards of the products provided, helping them continue as customers. Interaction in industry groups is also important to help all members of the group learn process safety lessons from each other, providing the opportunity to address similar potential problems.

The types of cultural interactions with stakeholder groups mirror closely those used in workforce involvement: *build trust, establish open and frank communication,* and then *understand and act on hazards/risks.*

The short-term effects of failure to conduct stakeholder outreach may not be noticeable. However, continued neglect can lead to serious problems. For example, neighbors and emergency responders may forget what to do in case of an offsite release. But even beyond process safety, failing to build trust and open communication with the community can eventually lead to resistance to the facility, making it much harder to expand and grow.

Contractor Management (Element 11)

Companies frequently use contractors in place of employees, for several reasons. Contractors can for example provide unique skills needed only occasionally and supplement staff during periods of high activity such as turnarounds. Contractor assignments may range from short assignments of hours or days to quite long-term. Contractors also provide professional services such as engineering, asset integrity inspections and expert consulting.

Contractors may be exposed to the same process safety risks as employees, and in some cases the risks may be greater. Therefore, any process safety risk management activities should control the risk that contractors face to the same level as employees. However, preparing the contractors to work safely and within the framework of the facility's PSMS can be complicated. Often, the terms of the contract specify that the facility cannot manage or train contractors directly, to avoid co-employment issues. In such cases, the facility must instruct the management of a contractors' company about hazards and protection measures, and then the contractors' management must instruct them. Similarly, providing corrective feedback may also have to go to the contractor through their management. Finally, contractors sometimes use sub-contractors, and this adds to the management challenge.

As challenging as managing contractors within the facility's PSMS can be, aligning contractors with the facility's culture can be much harder. Contractors arrive influenced by their own company's culture and the cultures of other facilities at which they've worked. They are also usually motivated by business commitments to provide services on a competitive schedule and price, enhanced by a fear that failure to meet commitments could hinder repeat business. As discussed in section 2.9, this environment is quite conducive to normalization of deviance.

They key to aligning contractors to the culture of the facility is a firm commitment that the imperative for process safety applies to contractors as well as employees. That commitment should be factored into the business arrangements with the contracting company and reinforced with the contractors and their management. Trusting relationships should be established with contractors and their management so that contractors feel comfortable providing open and frank feedback, and barriers to feedback should be avoided.

This approach will generally be more effective for longer term contractors. Shorter term contractors may not be present long enough to conform with the facility's culture. For such contractors, the first line of defense is the company's employees who work with the contractors. If they are working in a strong culture and truly empowered to fulfill their process safety responsibilities, they should be able to guide the contractor to the correct short-term behavior. As for longer-term contractors, the imperative for process safety must be clearly established, both in words and through the contract.

Conduct of Operations (Element 15)

Conduct of operations means execution of operational and management tasks in a deliberate and structured manner. It is also sometimes called "Operational discipline" or "Formality of operations." Conduct of operations relies heavily on every process safety core principle, and essentially operationalizes the process safety culture.

Conduct of operations applies to routine process operations and maintenance, as well as to non-routine operations including start-up, abnormal operations, and shut-down. Emergency shut-down has a special relevance to process safety culture that will be discussed.

Conduct of Operations also applies to all levels of leadership as well as the staff functions that support the PSMS. For example:

- Senior leaders need to conduct the corporate risk review process regularly and follow up to close gaps.
- Management needs to oversee all functions and hold them accountable for performing their specific roles in the PSMS.
- The HR department needs to consider process safety competencies when screening and hiring new employees.
- Engineering must follow the applicable standards and RAGAGEPS when designing and installing equipment.

Management and workers both have responsibilities for conduct of operations. Management defines procedures and standards, and controls for their consistent use. Workers commit to following the procedures and standards without variation by shift or unit. Both commit to performing their duties alertly, with due thought, full knowledge, sound judgment, and a strong sense of pride and accountability.

Signs of effective conduct of operations include:

- Consistent practice of established work processes and procedures, which are followed,
- Effective shift turnover practices,
- Consistent and proper use of safe work permits to control work,
- Effective and consistent use of interlocks, bypassed only with proper evaluation,
- Consistent use of bonding and grounding,
- Excellent general housekeeping,
- Few overdue action items; and
- No ad hoc trials or modifications.

Conduct of operations is clearly linked to combatting the normalization of deviance. It starts with an insistence that procedures should be followed. This must be supported with procedures that can easily be followed. Common problems with procedures include confusing format, language that is not easily understood by all operators, and specified actions that do not actually work as written. See Section 5.4 for more on operating procedures.

Once proper procedures are established, the role of leaders in conduct of operations is to ensure that they are followed consistently by all shifts. This means that leaders need to monitor performance in night and weekend shifts as well as day shifts and correct any differences found. Leaders also need to partner with workers to assure that shift turnover takes place sufficiently thorough that the new shift can be fully aware of the status of the process they are taking over and can continue without missing a step.

As part of conduct of operations, leaders should ensure that personnel are fit for duty. External factors such as fatigue, illness, distracting personal problems, and drugs and alcohol can affect an employee's ability to carry out their role. Special attention should be given to managing fatigue, since workers could be motivated by overtime pay, while management could be motivated by limiting numbers of full-time employees. Workers should be looking out for each other. A worker unfit-for-duty should not be allowed to work.

Emergency shut-down situations are a strong test for both conduct of operations and process safety culture. Operators should feel empowered to shut-down the process when they believe that conditions warrant. Indeed, they should feel obliged to do so. Similarly, technical resources need to have the same discipline when troubleshooting process problems and realize that they too have stop work authority.

To reach this state, leaders need to overcome some natural biases, their own as well as the workers'. Operators and leaders alike know that shut-downs can mean lost revenue. Operators need to put at ease that they will not face reprisals for shutting down. This should be frequently communicated through the leadership chain. Then, after shutdowns, leaders should acknowledge the correctness of the shutdown, while carefully avoiding any inadvertent signal that the shutdown was unfortunate or unnecessary.

Indeed, shutdowns may be, after investigation, found to have been unnecessary. This could lead operators to second-guess themselves, or believe they need higher level approval to shut down the next time. Operators may also come to view recovering from the shut-down – cleaning out equipment or performing a tedious start-up – as another disincentive to shut-down the next time. Leaders need to address this head-on by leaving no doubt that the operator's actions were commendable and correct.

Measurement and Metrics (Element 18)

Metrics, sometimes referred to as Key Performance Indicators (KPIs) are common management tools to monitor conditions and drive improvement. As such, metrics are important parts of both the PSMS and process safety culture. Leadership should therefore establish an appropriate number of metrics, tracking both the PSMS and the culture. In recent years, API (Ref 5.2) and CCPS (Ref 5.3) have collaborated to suggest useful leading and lagging metrics to use for these purposes.

In determining measurements and metrics, leaders should also consider how they will be used. In the words of an anonymous industry manager who said, "What gets measured can be corrupted." This reflects several ways that well-meaning metrics can possibly have effects other than those intended. For example:

- **Measuring near-misses:** A goal to reduce near-misses could lead to near-misses not being reported.
- **Measuring loss incidents:** An incentive for reducing actual incidents could lead to covering-up incidents.

- **Measuring task completion:** An incentive for completing all tasks could lead to checking-the-box instead of fully completing the tasks.

Both technical and human barriers to collecting metrics exist. First, metrics can be time-consuming to collect, process, and analyze. Often metrics are based on information collecting from several functions and systems. These systems may be based on different computer file formats and units that need to be translated, and then combined with data pulled from different systems in a time-consuming process. Second and perhaps more challenging, the people providing information these metrics may be concerned that they will be judged based on them.

Leaders need no prompting to pay attention to metrics that represent lower performance. However, leaders should also pay attention to metrics that change very little, especially indicating consistent performance, whether consistently good or consistently bad. In these cases, the effort to collect metrics that appear to provide no benefit could become resented and thereby undermine all metrics collection efforts. Conversely, consistently good metrics values could be the result of systemic problems. For example, on-time performance of maintenance tasks could be the result of regularly deferring the due dates, or excluding a problematic grouping of equipment from the data collection effort.

Measurement of the culture is also important. Unfortunately, culture meters have not been invented. Instead, culture is measured by a range of proprietary and public surveys of employees that are administered periodically. Appendix F of this book contains an extensive list of questions that can be selected from to evaluate the status of an organization in one or more of the cultural core principles.

Management Review and Continuous Improvement (Element 20)

CCPS (Ref 5.1) defines this element as "the routine evaluation of whether management systems are performing as intended and are producing the intended results as efficiently as possible." From a culture standpoint, Management Review is the ultimate confirmation to the entire organization that the corporate and site leadership is serious about their commitment to process safety, to learning and improving the process safety culture, and all the culture core principles in between.

Management review related to process safety should be performed at all levels of management, starting at the Board level and cascading through the organization to first line leaders. Leaders review with their groups and group members their individual and collective process safety goals, as they align with corporate goals. Management review should address topics such as:

- Gaps, and plans made to close them,
- Metrics trends and quality, and appropriate actions,
- Roadblocks to reaching goals or closing gaps,
- Improvement opportunities; and
- Resources and personal development needed.

With 20 elements in CCPS's Risk Based Process Safety and at least a dozen elements in any management system, it is unlikely that any leader will be able to cover every element in a single session. However, over time, each element should be addressed on a regular schedule, possibly addressing elements of greater importance to the group, facility, or company with greater frequency.

Some facilities include management review for process safety as a standalone activity, while others incorporate it into general broad-themed leadership team meetings. While the former may be more conducive to studying details, the latter has the advantage of treating process safety as part of the overall portfolio of business topics each group addresses. Both options are acceptable; the key is that management review happens.

5.2 RISK MANAGEMENT-RELATED ELEMENT GROUPING

All voluntary and regulatory approaches to managing process safety have some form of risk analysis and risk management as a central theme. At their core, these management systems seek to evaluate risk in some way, and to reduce any unacceptable risks "As Low as Reasonably Practical (ALARP)." The ALARP principle is explicit in several national regulations and to CCPS Risk Based Process Safety, and implicit in other regulations.

From the perspective of process safety culture, this grouping of elements drives how companies understand and act on hazards and risks.

Hazard Identification and Risk Analysis (Element 7)

The process of identifying hazards and analyzing risk is typically performed on every operating unit within a facility many times over its lifetime. The methods used may be tailored to the specific situation (Ref 5.4), but generally involve the following steps:

1. Identify the hazards of the process (e.g. toxicity, flammability, reactivity, etc.).
2. Estimate the potential consequences that could occur under process volumes and conditions
3. Identify the process deviations that could lead to these possible consequences.
4. Estimate the probability that these deviations could occur.
5. Identify the safeguards that prevent the consequences, and their probability of failure.
6. Determine the process risk.

Some unacceptable risks may be identified through this exercise. The hazard analysis team then identifies improved or additional independent safeguards, or consequence reduction measures, to reduce the risks to meet corporate risk criteria. Hazard analyses are typically repeated or revalidated every 5 years to check original assumptions and capture the impact of any changes to the process or process systems.

Hazard identification and risk analysis (HIRA) requires great attention to detail. However, it is often done under significant time pressure. It can be a key milestone in a process design or change, or as one of many such analyses that need to be done on a regular schedule. Under time pressure, PHA teams can make two possible errors:

- Work overtime trying to meet the deadline. This can lead to mental exhaustion and missing key hazards and safeguards by error; or
- Cut corners. This can lead to missing key hazards and safeguards through incomplete work.

For these reasons, leaders should take steps to alleviate excessive time pressure on HIRA teams, while taking steps to make sure their work is completed with the *appropriate sense of vulnerability*. Leaders should consider kicking off significant HIRA reviews, stressing the importance of the team's work and acknowledging the potential consequences to the facility if an incident happens. Leaders should also make sure that the HIRA team includes the diverse representation including process experts, HIRA experts, and operators, and that the team members have *mutual trust* and engage in *open and frank communication*.

Finally, leaders need to make sure that once the hazards and risks are *understood*, they are *acted on*. Action items from HIRA studies need to be fully addressed in a timely manner. Every action item does not need to be performed exactly as written by the HIRA team. Sometimes a better solution may be found after further study outside the HIRA meeting. Sometimes, the HIRA team may be found to have been too conservative, and a lesser solution is acceptable. Conversely, sometimes the recommendation may be found to not fully address the risk, and stronger measures are found to be needed. In the end, a recommendation should be considered closed only when the risk that the recommendation addressed has been managed by implementing a suitable solution.

In the Risk Based Process Safety approach as well as some corporate and regulatory approaches, the process risk may be used to guide the efficient and effective use of resources in carrying out the PSMS elements. From a culture perspective. tailoring level of effort to risk also helps *empower employees to fulfill their process safety responsibilities* by focusing their efforts productively. Table 5.2 provides examples of how higher and lower risks might be addressed in some PSMS elements. While Table 5.2 shows actions in two categories of risk, companies may have three or more action categories.

Table 5.2 Example of Tailoring PSMS Actions to Risk

PSMS Element	Higher Risk	Lower Risk
HIRA	Deeper risk analysis, e.g. QRA	Faster risk analysis, e.g. checklist
Asset Integrity	More rigorous inspection, testing, and preventative maintenance schedule	Run to failure
Management of Change	More rigorous evaluation; higher level sign-off	Less rigorous; lower level sign-off
Auditing	Supplement required audits with more frequent informal audits	Required audits only
Metrics, management review	Specific metrics, more frequent management review	General metrics, less frequent review

Asset Integrity (Element 10)

CCPS uses the term "Asset integrity" to more completely describe the process safety element commonly called "Mechanical integrity." However, mechanical integrity as defined in regulations commonly applies only to inspection, testing, and preventive maintenance activities related to a specific set of equipment specified by a regulation. A company with a strong process safety culture should gravitate more towards the holistic asset integrity approach rather than rely on the more limited scope of mechanical integrity. The US OSHA PSM regulation, which has also been adopted in whole or in concept in many other countries, provides a useful starting point, as do other country-specific regulatory approaches. The specification of recognized and generally accepted good engineering practices (RAGAGEP) can be a very useful tool to help companies get started.

Asset integrity addresses all equipment used in hazardous processes. Asset integrity also involves design activities such as material of construction choices and the design of the process and layout for maintainability. More information specifically describing asset integrity can be found in Reference 5.5.

The goal of asset integrity is to ensure that:

- Piping, vessels, and equipment safely contain the process,
- Instrumentation and control elements function as required; and
- Interlocks, relief systems, and safety instrumented systems perform their function when called on.

By doing so, the facility can help assure that the frequencies of equipment failures in the facility are no greater than what was assumed in the risk assessment. This helps keep the facility's risk within the company's risk criteria. When asset integrity is functioning well, inspections and testing will periodically reveal that equipment or components must be replaced. When a component is critically deficient, *strong leadership* should be demonstrated by having the component replaced promptly, rather than tolerating the *normalization of deviance.*

Asset integrity activities are constantly being carried out by the facility. Most of these activities are done by a separate maintenance organization. This organization typically interfaces with operations staff and with process safety specialists, but operates independently of either. Inspection and testing functions may be in yet another organization, and facilities may also have a separate reliability. Clearly, asset integrity cannot fulfill its role within the PSMS unless leaders *foster mutual trust* and *ensure open and frank communication* between all these functions.

When a facility's asset integrity element is functioning well, the primary role of the site leader is to ensure that *normalization of deviance* does not start to creep in. This may happen overtly and be seen by, for example, a rise in missed inspection deadlines. The signs may also be subtle and indirect, such as poor housekeeping and increased drips from process equipment and steam traps.

During periods of business slowdown, companies may consider whether asset integrity personnel can be reduced. To the degree that process units are idle or operating in a manner that requires less maintenance, reduction in asset integrity resources may be acceptable. The controlling factor, however, is not the production rate. It is the *imperative for process safety.* Asset integrity resources can be reduced only if sufficient resources remain to assure that failure frequencies remain at or below the failure probabilities determined in the risk analysis. In other words, reductions can be no more than needed to (*understand and*) *act on process hazards/risks.*

Asset integrity can also pose cultural challenges coming out of slowdowns and generally in any period of growth. As production rate increases, the resources required to maintain asset integrity will also increase. This may increase demands on existing staff, tempting them to skip steps and normalize deviance. Staff additions pose another challenge. New personnel must be brought into the culture and adopt it. Additional care should be taken that the new personnel do not bring negative cultural aspects from other places they have worked.

In recent years, asset integrity efforts have experienced number of challenges that have led to incidents. These include:

- Inferior castings, bolts, and equipment that contain voids, stresses, or other manufacturing defects but pass positive material identification,
- Asset integrity database errors introduced during asset integrity database management, upgrades, and migration,
- Components that are not tagged and therefore not included in the asset integrity database; and
- Neglecting to improve design and maintenance practices as they evolve in the industry, including useful information from outside the industry sector.

In a strong process safety culture, leaders empower the technical staff to study emerging issues that can improve the way they discharge their process safety responsibilities and defer to their expertise when they raise issues such as these.

As in most other PSMS elements, asset integrity can be threatened by time pressures. This particularly can be a challenge with asset integrity tasks that need to be done during a turnaround. Keeping turnaround as short as possible has significant competitive advantages. Nonetheless, leaders should maintain the imperative for process safety and defer to expertise before concluding the process can be restarted. However, hurrying to restart before critical asset integrity tasks have been completed, including removing blinds, replacing relief valves, and restoring bypassed interlocks, can be deadly.

Emergency Management (Element 16)

When process safety incidents occur, facility personnel should take actions that help reduce the consequences of the incidents. These actions include evacuation to a safe location, use of emergency masks, sheltering in place, first response, offensive response (e.g. to close an isolation valve), and firefighting, among others.

Since each emergency is different, it is impossible to develop specific procedures to address every scenario. Instead, specific emergency management personnel need to be expert at putting together the skills and resources at the disposal to effectively address the emergency. Everyone else at the site needs to be trained on a range of specific emergency management skills. Training should be done regularly, so everyone at the facility can carry out their role correctly and without delay.

In many plants, emergency management personnel may come from outside the plant. This can include industrial neighbors who partner with the facility in a mutual aid agreement as well as emergency responders from the local community. The cultural implications of these external stakeholders were discussed in section 5.1.

Emergency management can readily become subject to normalization of deviance. Since process safety incidents are infrequent, it can be easy to forget to plan, evaluate emergency procedures, and conduct drills. Ironically, the temptation to deviate from emergency preparedness could increase as culture and PSMS performance improves and incidents become even less frequent. However, emergency management is an integral part of risk management, and must be maintained, just as process equipment must be maintained. Culturally, emergency management should be treated as part of the imperative for process safety and monitored through the management review element (see section 5.1).

Two common culture challenges involve incipient incidents. The first involves discovery of a person found collapsed on the floor or inside equipment. People's natural motivation is to rush to their fallen colleague to render aid. The second involves taking heroic measures to stop a release, such as accessing a shut-off valve in the middle of a vapor cloud or liquid spill.

Both types of actions represent some of the same qualities desired in a strong process safety culture: a commitment and drive to protect co-workers and prevent major incidents. However, experience has shown that such actions typically fail to save the day, and the responder often becomes injured or dies. Instilling a strong sense of vulnerability and combatting the normalization of deviance (i.e. from the procedure that says, "evacuate immediately") can help prevent these accidents. Equally important is for leaders to show consistently that they care for the workers' safety. When workers feel that their leadership and coworkers have their backs, it should be easier for them to trust that evacuation is in fact the correct response.

Incident Investigation (Element 17)

Every incident and near miss represents one or more failed safeguards or a missed risk analysis scenario. While safeguards can fail randomly, they usually fail due to one or more failures in executing the PSMS. Leaders should therefore view any incident or near-miss as evidence of one or more PSMS deficiencies that could lead to other incidents on the site. Incident investigations that do not identify and address PSMS failures are an indication of a weak process safety culture. By finding these deficiencies and correcting them, future incidents can be prevented. Additionally, incidents and near misses should heighten the sense of vulnerability.

Near-misses are particularly valuable, in that they highlight the deficiencies and vulnerabilities without causing serious consequences. Leaders should therefore instill mutual trust that reporting near misses will be welcomed and acted upon, and should ensure open and frank communications to remove barriers to reporting.

Certainly, investigations of incidents and near misses consume time and resources, and may delay the restart of operations. However, if the investigation finds root causes in the management system and culture that can be corrected, future incidents – with all accompanying costs and delays – can be avoided. Ultimately, investigating and then following up on findings and recommendations will strengthen the management system and promote learning to advance the culture.

As of this writing, there remain companies, industry sectors, and countries that view incident investigation in ways that are negative to process safety culture. One harmful view is that an incident investigation should be only cursory or else it will identify shortcomings that could be targeted by regulators and attorneys. This certainly can happen. However, regulators and opposing attorneys can also conduct their own investigations and draw their own conclusions. In other words, there is no real benefit to this approach, and real opportunities to advance the culture and the management system are lost.

Another harmful view is of incident investigation as a tool to assess blame on the operator or mechanic whose error caused the incident. By doing so, management motivates workers to hide near-misses and cover-up incidents. Communication of bad news up the chain of command becomes stilled and opportunities to prevent future incidents are lost. Finding blame should never be the objective of an incident investigation.

That is not to say that a proper investigation will never find that a worker or manager acted counter to the company's performance policy or broke a law. This is discovered from time to time, and when it does, the imperative for process safety requires that the individual receive the appropriate discipline.

However, the investigation should not stop there, and instead continue until the root causes are identified, including the cause for why the illegal or anti-policy act had not been detected and prevented. Indeed, if such acts were committed, the trust that management will properly address safety problems can be broken.

Auditing (Element 19)

Like audits of any other business practice, PSMS audits serve critical roles in governance and risk management. Process safety audits are independent reviews to determine if PSMSs are functioning as intended to manage process risks and to comply with regulations and corporate standards. Companies and facilities with a strong process safety culture will also use audits to identify opportunities to improve the PSMS. Audits are typically conducted every 5 years, although high-risk facilities may be audited more frequently.

Audits also provide a window into the process safety culture of the organization. It is possible, and indeed a good practice, to audit process safety culture specifically. Appendix F provides a list of sample questions that can be incorporated into a culture audit.

Audit findings describe the non-conformances with regulations and standards identified. Some companies ask auditors to recommend means to close conformance gaps, while others prefer auditors to focus only on auditing. The choice of approach depends partly on the company's legal philosophy and partly on the strength of the culture. In general, if the company has a strong process safety culture, either approach can be successful. However, if the culture is not yet strong, the auditors should not offer recommendations. This often leads to cosmetic solutions that aim to reduce the number of findings, but that do not fully close the gap.

Facilities with strong process culture welcome audits and encourage their personnel to cooperate fully with auditors.

Likewise, auditors perform their role in a constructive way, showing the facility personnel that the audit is a constructive activity, not punitive. Fostering mutual trust helps make the audit as effective as possible. Facility personnel attempting to guide auditors away from certain areas or records (presumably because findings would be likely) is a clear indicator of a weak culture.

Disagreements between the facility and auditors over findings are common. These can arise due to:

- Information or documentations missed by the auditor, or not known by the interviewee,
- Differences in interpretation of a regulation or standard,
- Differences in engineering approaches,
- Findings by the audit team that had not been identified at a recent regulatory inspection;

and many others. Disagreements can generally be resolved to the mutual satisfaction of the facility and the auditors. When they cannot be resolved, the parties agree to disagree, and the item should be recorded by a finding, which can be resolved later by those with relevant experience.

Since audit findings represent weaknesses that could lead to accidents, facilities with strong process safety cultures strive to correct the findings as quickly as possible. A finding that repeats from a previous audit indicates a cultural weakness.

Signs of a weak culture can be detected readily by the audit team. These include:

- Fear of the audit.
- Key personnel missing during opening and closing meetings.
- Difficulty meeting with the management team and key personnel, e.g. for interviews and daily review meetings.
- Cleaning up behind the audit team to remove findings before the audit is finished.

- Evidence of a check-the-box mentality, such as checklists that appear to have been completed without thought (or photocopies of the checklist with only the date changed).
- Poor housekeeping and disorganized filing.
- Subtle or unsubtle attempts to guide auditors away from certain areas or files, or to reduce auditors' time on site (e.g. delays in accessing site, offers to take auditors out for long lunches or early dinners).
- Negotiating findings based on cost (e.g. "That would put us out of business").
- Challenging every finding.

5.3 PROCESS-RELATED ELEMENT GROUPING

The previous section discussed managing the risk of the process. This section addresses the process itself. The process safety elements in this section address the management of the process: the standards used in designing and operating the process, what is known about the process, how the process is changed, and how it is started up.

While largely a knowledge-based grouping of elements, culture still plays a strong role, as will be discussed in this section.

Compliance with Standards (Element 2)

In the mid- to late 1800's as steam became increasingly used for propulsion, heating, and processing, many steam boilers suffered explosions, causing great damage and injuring many workers. Engineers came together to develop standards for the safe use of steam boilers. Use of these standards helped reduce significantly the number of boiler explosions.

Since that time, many standards have been developed. Hundreds of organizations around the world develop and refine standards applying to an ever-increasing number of equipment, process technology, and process design. Standards are also often developed by companies to focus on practices specific to that company. Regulations are also considered by this element. While not strictly standards, process safety regulations tend to be developed and written as if they are standards, and some are even referred to by that name.

Espousing a strong system of internal and industry standards provides a framework for correct design, proper installation, and effective inspection, testing and maintenance. This is indicative of positive safety cultures. The absence of standards, or inconsistent application of standards is conversely an indicator of a weak culture. If equipment is not designed to a formal code or standard, how will you know it will meet the process demands or how to properly insect and test the equipment?

By determining which standards are applicable and understanding how they work, companies can leverage experience in designing and operating processes and management systems, while complying with applicable regulations. Companies should keep abreast of new developments in standards and regulations, and address changes appropriately in its technology and management systems. Additionally, standards developed for different industry sectors or even countries should be considered if they address challenges in the facility's sector. The American Petroleum Institute's RP-755 addressing fatigue management is an excellent example of a standard with broad usefulness outside the US petroleum sector.

Establishing and maintaining internal corporate standards is an effective way to keep abreast with developments in standards and interpret these standards applicable to company technology and culture. Some companies will implement internal standards at the facility level to explicitly address local and national standards. Others will strive to have one set of corporate standards that applies regardless of location. This is largely a matter of preference. The important thing is that it is considered part of the imperative for process safety to identify, understand, and implement the applicable standards.

Some debate exists on the applicability of newer versions of standards to processes and equipment that was designed by older versions. Some standards (and regulations) provide grandfathering provisions, e.g. accepting as compliant equipment that complied with the older version that existed when the process was built. From a process safety culture perspective, this debate is academic. In a strong process safety culture, leaders should understand and act upon the hazards and risks that the updated standard seeks to address. This could mean modifying the process or equipment to comply with the updated standard. Alternatively, some other risk reduction measures may be taken to allow the process to meet the company's risk criteria.

Cultural problems almost certainly exist when a facility or company bases their compliance with standards activity solely on regulations. This may signal culture gaps in deferring to expertise (including external expertise), understanding and acting on hazards and risks, and indeed in the imperative for process safety.

Process Knowledge Management (Element 6)

Process knowledge (also commonly referred to as process safety information or PSI) is the written body of information that describes how the facility was designed, built, and how it is operated and maintained. CCPS (Ref 5.1) refers to knowledge rather than information to highlight the value of understanding all the information that is collected, rather than simply collecting it.

Process knowledge serves as the basis for other PSMS elements, most notably HIRA/PHA, operating procedures, MOC, and asset integrity. Therefore, whenever changes are made (e.g. via the MOC element), the process knowledge must be kept updated. Without this, the stage becomes set for normalization of deviance problems and errors, such as:

- Missed consequence scenarios and incorrect risk reduction determination in process hazard analyses,
- Errors in operating procedures,

- Setting or designing a relief system incorrectly,
- Incomplete isolation and lockout/tagout,
- Failure to inspect, test, or maintain critical equipment,
- Failure to detect a problem during an audit; and
- Increased difficulty in determining incident root causes.

Keeping process knowledge up to date should be an explicit part of the *imperative for process safety*. Whenever a process is changed, updating the process knowledge generally happens after the process was restarted and is operating smoothly. This is practical, as some minor changes may have been made in the field compared to the original proposed change, either before or after start-up. However, at this stage, pressures to move on to the next improvement can distract engineers from updating the process knowledge. Therefore, updating of the process knowledge should be the last step before a management of change process is considered closed. Leaders should guard against *normalization of deviance* by validating this vital step has been completed.

Process knowledge should be collected in a system that makes it easy to locate the needed information, while providing security to guard it against becoming inadvertently corrupted. Some facilities, meaning well, make the process knowledge database accessible only by the process safety organization, arguing that this allows them to better control management of change and safe work practices. However, part of the function of process knowledge management is to provide the knowledge to those in the facility who need it. In an organization with a positive process safety culture, access to pertinent information should be permitted to anyone who is *empowered to fulfill the relevant process safety responsibilities*.

Management of Change (Element 13)

Facilities continually seek to improve processes to improve cost, productivity, and quality, as well as environmental and safety performance. As discussed throughout this book, the mindset that drives this quest for continual improvement can be susceptible to *normalization of deviance.*

MOC is essential to manage ongoing process improvement safely. MOC makes sure that changes do not introduce new hazards, that the process continues to meet the company's engineering standards and risk criteria, and that all elements of the PSMS are updated to reflect the changes.

In a strong process safety culture, MOC should be considered an essential and valuable activity. MOC helps ensure that changes to processes and personnel do not inadvertently introduce new hazards or increase process safety risk. MOC includes a formal review and authorization process that evaluates changes to equipment, process conditions, procedures, and organization, addresses any needs to improve safeguards, and assures that procedures, asset integrity, and process knowledge is updated to reflect the change.

The MOC procedure should be applied for all changes that are not replacements-in-kind. A replacement-in-kind is a new component, material, or person that meets the same specifications as the original. Companies with a strong process safety culture will nonetheless consider replacements-in-kind carefully, to make sure that the original specification was sufficiently complete. Sometimes, new raw materials and process components may contain defects or impurities that are not addressed in the specification, but can cause process failures. For example, incidents have resulted from castings sourced from countries with emerging economies that were identified as replacements-in-kind based on metallurgy but contained voids that led to premature failure.

Leaders should design and enforce use of the MOC procedure as part of the *imperative for process safety*. This should be done in such a way that all employees come to appreciate the importance of MOC and perform it with an appropriate *sense of vulnerability*, *understanding and acting on hazards and risks.* As part of accomplishing this, the MOC procedure should consider several points:

- Changes should not be made without MOC.
- Replacements-in-kind should be carefully evaluated to ensure they are not actually changes.
- The requester should include MOC in the project timeline.
- The requester should provide a complete description of the change using information from the process knowledge management system.
- The level of MOC evaluation should be based on process risk, with higher risk processes subject to more rigorous MOC reviews.
- At no time should MOC be rushed or treated as a check-the-box exercise.
- Similarly, management and personnel should avoid pressuring reviewers to approve an MOC before a sufficiently thorough evaluation has been done. Reviewers should resist such pressure and provide the appropriate *open and frank communication* if they feel undue pressure.
- The level of MOC approval should be based on process risk, with higher risk processes requiring approval by higher levels in the organization.
- Conflict of interest should be avoided. The persons requesting the MOC and sponsoring the change should not be approvers.
- All action items identified in the MOC should be closed-out and verified in the field.
- After the change has been implemented, the requester should update the information in the process knowledge management system.

In recent years, many facilities have sought to create MOC efficiencies. Electronic MOC systems (e-MOC) have become common. These can help address the document routing and document management needs and can help expedite the procedure through workflow management. A potential downside of E-MOC is that it can seduce participants to act on their own rather than meeting with the full group of individuals involved in the MOC, thereby reducing open and frank communication. Therefore, leaders should make a special effort to encourage communication in the MOC process.

E-MOC system make it easier to proliferate the number of approvers. More is not always better. When too many approvers are listed, each may think that another one will catch any errors, so they give the MOC a cursory review and approve it. If all approvers take this approach, important issues will be missed.

The MOC for process safety can be combined with the change management systems for other considerations, such as quality and environment. This can make good sense for purposes of efficiency and for cross-fertilization of ideas in the MOC reviews. Care should be taken when doing this that process safety does not get lumped into occupational safety, for the many reasons discussed throughout this book.

Emergency MOCs may be required from time to time to keep the facility running when some component fails. With 168 hours in a week during which approvers are likely to be in the plant less than 60 hours, emergency MOCs are likely to occur on off-shifts and require verbal approval. In a strong process safety culture, emergency MOCs should be rare, and occur only when the risk of not making the emergency change outweighs the risk of making the change. If approval was given verbally, the proper documentation of the verbal approval should be done as soon as the approver returns to work. However, when emergency MOC is implemented, the process should be returned to its original state as soon as possible.

Temporary MOCs may also be required from time to time for product or process trials or repairs. Temporary MOCs should be planned and scheduled, and should not be conducted on an emergency basis. The trial should have an expiration date, after which the process is returned to its normal state. From time to time, trials may be conducted with the intention of implementing the trialed condition permanently if the trial is successful. Facilities with a strong culture should realize that the temporary MOC may be approved with less robust safeguards than would be accepted for a permanent change. Therefore, before making the temporary change permanent, a formal MOC should be conducted to evaluate the long-term implications of the change.

Facilities with strong cultures recognize that changes to personnel can affect process performance as well as process safety culture. Such facilities should also have an organizational MOC system (OMOC) to review personnel changes, including ensuring that:

- Any employee (including leaders) taking on a new assignment has the skills and capabilities specified to perform the process safety requirements of that role; and
- During restructuring and periods of high activity, critical roles are not left inadequately addressed

In performing MOCs, facilities with strong process safety culture remain aware of the potential impact of creeping change. This can occur as the process changes over time, eventually migrating far enough from the original operating window that the process knowledge no longer is sufficient to allow proper identification of hazards and evaluation of risk reduction measures. MOC reviewers with a strong sense of vulnerability should look out for the impact of creeping change, and require additional process data be provided if needed to assess the impact of the change.

Operational Readiness (Element 14)

Many PSMSs incorporate a pre-startup safety review (PSSR) element to ensure that processes are fully ready to be started up and operated safety. When CCPS created the Risk Based Process

Safety management system, it recognized that the term PSSR is defined in the OSHA PSM regulation too narrowly for a facility with *an imperative for safety*. Therefore, while PSSR applies only to new and changed processes, operational readiness (OR) also addresses startup after:

- Maintenance or turnaround,
- Idling,
- Precautionary shutdown (e.g. for impending severe weather),
- Short-term shutdown; and
- Longer-term mothballing.

Experience has shown that the frequency of incidents is higher during process transitions, particularly startups. Blinds need to be removed, piping must be complete and leak-tight, valves need to be lined up properly and, if automatic, switched to the correct control positions. Controls and relief systems must be functioning and online. Utility systems must be operating adequately. The appropriate personnel must be trained and available to perform the required work. And the list goes on.

Many companies combine their systems for MOC and OR/PSSR. This can help assure that all aspects of changes are addressed and assist with regulatory compliance. However, this tends to leave out operational readiness activities related to any kind of restart that does not involve a change. Compounding the risk during restart is the sense of urgency, whether real or imagined, to get the systems up and running. Cultures that *maintain a sense of vulnerability* assure all processes are operationally ready no matter the length of the process interruption.

Like in many Risk Based Process Safety elements, the depth of operational readiness review may be based on the process safety risk. However, startup may impose additional risk compared to normal operations. This could be due to the operating conditions during start up, or due to the conditions the process was exposed to during the shutdown. For example:

- If the shutdown was due to severe weather, equipment may have been damaged by high winds or blowing debris,
- If the shutdown involved a flood, water may have intruded into the process and weakened foundations; or
- If the shutdown was extended, air may have intruded into processes that should be inerted. Process materials left in place may have changed composition.

Leaders should pay attention to such factors and increase the level of operational readiness review and approval required as appropriate.

Operational readiness reviews are typically the last thing done before start-up. This can put reviewers and approvers under significant pressure to complete their work. In weak cultures, approvers may also be put under pressure to allow start-up before they are satisfied that the process is safety to run. In strong cultures, leaders should *empower* the reviewers *to fulfill their process safety responsibilities* and *defer to the expertise of reviewers and approvers*. Likewise, leaders should not allow operational readiness to become a check-the-box activity, *preventing normalization of deviance*. Finally, as appropriate in all other PSMS elements involving review and approval, conflicts of interest should be avoided; reviewers and approvers should be different people than those responsible for the change and for the startup.

5.4 WORKER-RELATED ELEMENT GROUPING

If leaders create the environment in which strong culture can flourish, it is the people who are in the line of command and the support staff who implement that culture and live within it. That does not exclude leaders from this category. Leaders are workers too!

This group of elements addresses employees' capabilities required for them to be *empowered to successfully fulfill their process safety responsibilities*, the training required for them to assure those capabilities, and the work they do to safely operate and maintain the process. It should be clear that senior leaders have process safety competency requirements for their own positions and require training to help them maintain those competencies. Likewise, leaders should assure their teams competencies are maintained and developed.

Competency (Element 3)

Simply stated, a company should maintain sufficient competency to safely manage the hazards of its processes. Competency means knowing what skills and resources are required to operate safely and then providing those skills and resources. To accomplish this, leaders should also ensure that the organization has properly identified the required skills and abilities of each employee at each level, along with required skills and abilities of contractors if they are used.

Competency is not a once-and-done activity. Processes change, organizational structures change, technology advances, and skills can fade if not practiced regularly. People also change jobs and advance through the organization. Therefore, leaders should have an inventory of required skills and competencies, reevaluate competency requirements regularly, and develop employees' skills and abilities to allow them to meet the changing competency requirements of the changing workforce.

In a strong process safety culture, leaders know that the more competent the organization, the more efficiently it should be able to manage its hazards. Leaders should therefore invest in competency development for themselves and employees as part of *learning to advance the culture*. As leaders foster competency development, the organization will develop a questioning and learning environment. This can help *ensure open and frank communication*, particularly in traditionally hierarchical cultures where asking questions for the purpose of learning is an appropriate way to initiate communication.

Key employees in PSMS roles should not restrict competency building to internal development efforts. Since process safety incidents tend to be rare events, it is important for process safety personnel to participate in local, national, or global industry or technical organizations, meetings and conferences. This gives them direct access to lessons learned from other companies across the industry. A person's access to lessons learned is greatly facilitated by sharing their own lessons learned, Doing so *fosters mutual trust*. Some companies may be uncomfortable with this level of sharing. However, the value of sharing is so high that companies should find ways to appropriately manage the details of what is shared while enabling the exchange of lessons-learned.

There is ongoing debate whether senior facility and corporate leaders should have experience in the processes and technologies that they manage. The debate addresses other technical and managerial disciplines beyond process safety. The school of thought embraced by followers of the HRO approach (Appendix D) believe that technical competence in the discipline is essential, particularly when it comes to preventing catastrophic incidents. For example, in the nuclear industries of many companies, facility managers must spend a minimum amount of time in a nuclear safety role. The other school of thought believes that leaders do not need to have had experience in the discipline; they need only surround themselves with staff having the necessary competency.

Certainly, having the technical expertise helps, particularly in *preventing the normalization of deviance*. However, either approach can work, from a process safety culture perspective. The bottom line, with or without technical experience, leaders should:

- Understand the PSMS and its underlying principles,
- Know the hazards their organization is managing,

- Know the safeguards that protect the facility, its employees, and its neighbors from those hazards,
- Have a system to know that these safeguards are being maintained effective,
- Know enough about the technology to understand what they are approving and what they are asking their team and employees to do; and
- Know who has the technical knowledge for consultation when difficult questions arise.

<u>Training and Performance Assurance (Element 12)</u>

Training is the practical instruction in job and task requirements and methods. Training helps build the skills and abilities that individuals need to perform their jobs or prepare for new jobs. The skills and abilities for which training is needed for a given position are identified through the competency element just discussed. The training element also includes performance assurance, to confirm that training successfully imparted the required skills, leading to competency. This relationship is illustrated in Figure 5.2.

Figure 5.2 Relationship between training and competency

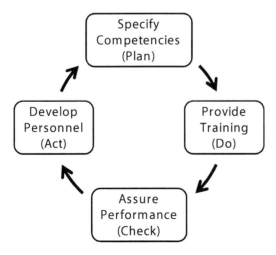

From the perspective of process safety culture, training provides the skills that give leaders the ability to *empower individuals to successfully fulfill their process safety responsibilities*, and the confidence that they may *defer to expertise*. Training also provides opportunities for leaders to reinforce *the imperative for process safety* and *maintain the sense of vulnerability*.

Every employee and contractor at a facility requires some form of process safety training. The need to train operators, mechanics, supervisors, and other production personnel should be clear. But even if the individual never works outside of the administrative office, they still need to be skilled in the necessary emergency procedures and understand the hazards managed at the location. Some office workers may require more training. For example, procurement professionals need to understand the process safety implications of changes in the sources of spare parts, replacement equipment, and raw materials.

Companies manage process safety training initiatives differently. It is not unusual, for a common training group to organize and manage training in all skill areas. Other approaches include managing all skills by department, and managing process skills through the process safety function. Each has benefits and potential drawbacks, but with strong culture, *leadership*, and *open and frank communication*, all approaches can work.

If training is managed outside of the sphere of process safety culture, it may be necessary to harmonize the training culture with the process safety culture. For example, if the overall training culture focuses on checking the boxes of required training – ethics, non-harassment, and so on – it will take some effort to establish a process safety training effort that focuses on competency and culture. That effort is necessary to prevent the overall training culture from undermining the process safety culture.

Computer-based training (CBT) is now common. CBT provides many advantages in terms of efficiently getting training to those who need it, tracking training, and conducting testing as part of the performance assurance activities. Leaders should be aware of the drawbacks of CBT, most notably that if a trainee does not understand some part of the training, there is no instructor to ask for clarification. CBT is also less useful for training that needs to be conducted hands-on, such as performing physical tasks like maintenance, inspection, and worksite evaluations. If CBT is used for such tasks, it should only be to provide basic familiarity, and be supplemented with in-person instruction and demonstrated proficiency.

A recent advance in CBT for process facilities is the use of simulators. These can be particularly useful for training operators on the processes they run. Various deviations can be imposed on the simulation, and the operator can gain experience in how to handle them. Simulation can also help trainees develop a *sense of vulnerability* by being allowed to virtually blow up the plant.

Whether training is in-person or CBT, it should try to incorporate hands-on elements. This could involve group or individual exercises, supervised work in the field, and simulators. Physically performing tasks helps people remember what they learned.

In certain topics, training cannot cover every eventuality. While some parts can be learned by rote, other parts require the trainee to develop understanding. For example, when training a supervisor how to prepare a safe work permit, the mechanics of filling out the permit and filing it can be learned by rote. However, the ability to recognize hazards and determine the appropriate safeguards requires developing deeper understanding.

The ultimate aim of training is proficiency. It is not acceptable for a mechanic or operator to perform their jobs correctly most of the time. Therefore, the target score is 100% for every training event and every trainee. The instructor should prepare and deliver the training materials, and follow up with the trainees sufficiently that each can achieve 100% on the final test. There are three measures of training success:

- Did the trainee completely learn the material (e.g. achieve 100% on the final test)?
- Did the trainee remember the material later, for example the following month?
- Did the training make a difference, i.e. was the trainee performing proficiently sometime later?

In addition to training required for the job role and the process, training may also result from actions generated in the PSMS. This can include training on process changes originated through MOC or new hazards or new safeguards arising from an updated risk analysis. There could also be new training precipitated by an incident investigation: for the process, the site, or elsewhere in the company or industry. If training is managed outside the process safety function, diligence is needed to ensure that training driven from such sources becomes part of the training for future operators.

Regulatory inspections around the world have shown that training is one of the PSMS elements in which there are the most gaps. To develop and maintain a strong process safety culture, training gaps cannot be permitted. Training must enable the successful behaviors desired in a strong process safety culture – *the sense of vulnerability, understanding and acting on hazards and risks*, being *empowered to fulfill process safety responsibilities*, and being qualified to be *deferred to for expertise*.

Operating Procedures (Element 8)

Operating procedures (OPs)are written instructions (including electronic) that specify the steps for a given task and describe the way these steps should be performed. Operating procedures should exist for steady state situations such as continuous processes as well as to transient situations such as startups, shutdowns, and batch processes.

Good OPs also describe the process, identify process hazards, and describe the measures required to safeguard against those hazards, not only process safety but also environmental, health, and occupational safety. Good OPs also describe safe operating limits, a troubleshooting guide, and emergency actions. Consequences of deviating beyond safe operating limits must be included to maintain the *sense of vulnerability*.

Operators must be trained on a procedure before they follow it in the field, and they must understand everything contained in the OPs. Following training, performance assurance should be done to make sure the operator follows and continues to understand the procedure. Ideally, training on procedures should be based on the procedure document itself, rather than on separate training materials. When it is necessary to have separate training materials, and especially when the separate training materials are being used in lieu of the procedure, this is a warning sign that the procedure is not adequate.

The necessity of following procedures was discussed in section 5.1 (Conduct of operations). This means also that procedures should be written so that they match what operators actually do. Moreover, procedures should be written in plain language, written to a comprehension level of no higher than 8th grade. Text should be well spaced with a line length no longer than the text on this page. Tables, figures, and illustrations should be provided as needed to enhance communication. All these measures will help operators follow the procedures and resist their temptation to stray from the procedure, leading to *normalization of deviance*.

Operating procedures should be controlled documents that are kept up to date whenever there is a change. Operators must use the current OP, and no old versions should exist except in the document management system. Most changes come through the

MOC process, and updating the procedures should be a requirement to close out all MOCs. Sometimes procedures may be changed for other reasons, such as new process knowledge or improvement in clarity. Such changes should also be tracked and controlled.

To verify consistent use of procedures, they should be reviewed periodically for accuracy and to verify that they are being followed. Some regulations and standards specify that procedures be reviewed every 1-3 years. In the absence of such requirements, facilities should set review cycle based on risk; review higher risk processes more frequently. These measures help *prevent normalization of deviance*.

Having operators or maintenance personnel review and comment on the procedures they use is another way to reinforce PS culture. When management acknowledges the value of these reviews, *it fosters open communication and builds mutual trust.*

CCPS (Ref 5.6) describes several ways of formatting procedures, each with advantages and disadvantages. Companies should select the style(s) that work best for them. Once selected, operating procedures should follow the standard style. Operators should be closely involved, both in selecting the standard style and in writing. This helps ensure that procedures are understood and followed. In appropriate cases, experienced operators could write operating procedures to be checked by engineers.

Experienced operators may come to know procedures so well they've memorized them. In good cultures, operators follow the procedures even if they know them very well. There are two reasons:

- If operators read procedures as they operate the process, they are less likely to normalize deviance or have an error of memory.
- When procedures change, memory will no longer be correct.

Therefore, leaders should be alert for of signs of memorizing procedures, such as failing to check off necessary steps and taunting of employees who do not or cannot conduct the procedures from memory.

Safe Work Practices (Element 9)

Non-routine work often increases risk and has directly caused catastrophic accidents. The safe work practices element controls non-routine work such as hot work, lockout/tagout, line-breaking, vessel-opening, confined space entry, and similar operations. The term control of work may also be used to describe this collection of activities. Some facilities will also include occupational safety practices such as fall protection, electrical safety in safe work practices. This is both convenient and acceptable.

In recent years, many companies have created inviolable rules for safe work practices, stressing the *imperative for process safety* and intolerance for *normalization of deviance*. These may also be termed "Life Saving Rules" or "Cardinal Rules." Such rules are part of the employee code of conduct, such that a violation leads to discipline up to and including termination. Life Saving Rules apply to ALL employees, regardless of tenure or position.

Based on successful industry experience, this approach should be considered as part of strengthening culture, as soon as the company is prepared to *provide strong leadership* in process safety. If such rules are adopted, it is important that the rules be enforced fairly and consistently, and not permitted to become a slogan with no teeth.

The use of written work permits should, and almost always does accompany the implementation of safe work practices. The permitting process starts with a job safety analysis, an exercise of *understanding and acting on hazards and risk* that is performed with a *sense of vulnerability*, rather than a purely administrative exercise.

As discussed above for MOC and OR, individuals engaged in the permitting process may be under pressure to complete their approvals quickly, so the desired work can get started. Leaders should make it clear that work permit writers should get the time they need to complete the job safety analysis properly, and workers should be given the time to perform their work safely. The safety of the job should be valued more than the act of completing the permit.

Good cultural practices for safe work practices include:

- Permit approvers should be well trained in hazard recognition and control of hazardous energy.
- Work permits should only be signed in the field, only after performing a thorough job safety analysis. Permits should never be signed in the office.
- Closure of the permit should also occur at the job site.
- Work should be scheduled as much as possible, to avoid everyone wanting a permit on Monday morning. If many people are simultaneously trying to obtain permits, this should not affect the thoroughness of the job safety analysis.
- Physical work should never begin before the permit is issued, nor should the permit be written and approved after the work is complete.
- Equipment should not be taken out of service without approval of the process operators.
- Permits should be accurate, reflecting the actual work to be done and the correct names of the workers. Each of the workers should personally attach their locks.
- The scope of each permit should be clearly and precisely stated. If the scope of work needs to change, a new or modified permit is needed.

In a strong process safety culture, every permitting exercise should be taken with fresh eyes. Every job is a little bit different, and field conditions may not be as expected. Every permit review and approval should be conducted with the same diligence the approver would use if they were going to have a family member perform the task. Anyone involved in the permitting process or in conducting the work itself should be able to feel secure in voicing objections or pointing out potential risks or flaws in job preparation activities.

Leaders should use audits and informal walk-throughs to verify that the safe work practices element is functioning correctly. Field visits also important to allow leaders to correct any errors, reinforce good behaviors, and identify improvement opportunities.

5.5 REFERENCES

5.1 Center for Chemical Process Safety (CCPS), *Guidelines for Risk Based Process Safety*, American Institute of Chemical Engineers, 2007.

5.2 American National Standards Institute/American Petroleum Institute, *Process Safety Performance Indicators for the Petroleum and Petrochemical Industries*, ANSI/API RP-754, 1st Ed, 2010.

5.3 Center for Chemical Process Safety, Process Safety Leading and Lagging Metrics …You Don't Improve What You Don't Measure, American Institute for Chemical Engineers, 2011

5.4 Center for Chemical Process Safety (CCPS), *Hazard Evaluation Procedures, 3rd Ed.*, American Institute of Chemical Engineers, 2007.

5.5 Center for Chemical Process Safety (CCPS), *Guidelines for Asset Integrity Management*, American Institute of Chemical Engineers, 2016.

5.6 Center for Chemical Process Safety (CCPS), *Guidelines for Writing Effective Operating and Maintenance Procedures*, American Institute of Chemical Engineers, 1996.

6
WHERE DO YOU START?

6.1 INTRODUCTION

Evaluating and then modifying the process safety culture of your facility or company can be a daunting venture, particularly if the required culture change is significant. This chapter addresses how to get started and provides a roadmap for your culture journey.

First, you should acknowledge that there may be existing or developing weaknesses in the process safety culture. Even the companies with the best performance in process safety have some weaknesses or have the potential to develop them. A company that denies that it may have weaknesses in its process safety culture has at least one weakness – a decreased *sense of vulnerability* – and probably other weaknesses as well. In stronger cultures, the feeling that weaknesses have all been corrected indicates a sense of complacency that can quickly compound via *normalization of deviance*. In less-developed cultures, denial may be based on a false sense of security taken from the wrong metrics or a focus on compliance, indicating a weak *imperative for process safety*. Therefore, all companies should search for cultural weaknesses, regardless of where they are on their process safety journey.

Making the case for culture change can be challenging. Marshall the facts carefully to show how process safety supports business and financial success, and how improving process safety

Essential Practices for Creating, Strengthening, and Sustaining Process Safety Culture, First Edition. CCPS. © 2018 AIChE. Published 2018 by John Wiley & Sons, Inc.

culture supports both. Support your case with strong examples, including near-misses that could have caused significant property loss and casualties. The free publication *The Business Case for Process Safety* (CCPS 6.1) may help you argue the financial case. Selected questions from Appendix F may prove useful in highlighting culture gaps that can resonate with leadership and enable some early successes. You may need to make your case at many levels of the organization.

6.2 ASSESS THE ORGANIZATION'S PROCESS SAFETY CULTURE

Before any changes or improvements to the process safety culture can occur at a facility, the existing culture must be assessed. This assessment should take place prior to the initial culture improvement effort, and before subsequent improvement cycles.

Ideally, the assessment should address all culture core principles. If budget or staff is limited, triage of the principles may be performed to narrow the focus of the assessment to those that may provide the greatest improvement for the least cost and effort.

Since most evidence of process safety culture exists outside of hard operational and financial data, assessing the culture largely consists of interviewing, observing, and surveying people as they go about their duties. To obtain an accurate picture, a large and diverse set of employees must be interviewed, from senior leadership through middle management and supervision to hourly personnel. Responses and observations will likely differ by level, reflecting a diverse set of opinions on the status of the culture. Understanding the differences in perspective by level can be as informative as the individual responses.

Culture assessments should consider the components highlighted in figure 6.1and discussed in the following paragraphs. The master culture assessment protocol and follow-up questions presented in Appendix F can be used as a starting point for tailoring your protocol. This protocol incorporates significant thought from prior work (Refs 6.2 through 6.9), as well as the experience of subcommittee members.

Figure 6.1 Components of a Culture Assessment

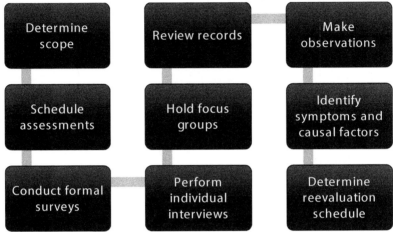

Determine Scope

If not already done, leaders should define the scope of the assessment. If the culture effort is company-wide, should the assessment also be done company-wide? Or, should it be focused on the site or sites perceived to need the most improvement, on those most welcoming to the culture effort? Should it focus on all the culture core principles, or just a few where gaps are perceived? Similarly, if the effort is site-wide, should the assessments also be site-wide or focused on a few units or departments?

There is no single right answer. Some companies may wish to start with a few model sites or model units to gain experience with assessments and fine tune them. After gaining experience and showing valuable results, assessments can spread more broadly. Some companies may wish to conduct assessments more broadly, so that all sites and units feel equally included in the process and change happens more quickly.

Culture surveys should be performed anonymously, and ideally by an independent party. At the outset of culture improvement efforts, mutual trust may not yet have been developed, so respondents may hesitate to give full open and frank input to assessors who represent company management. It may not be necessary to use an independent party for re-surveys of organizations where mutual trust is already high, but is still a good idea just in case there has been some slippage in the culture. Whoever conducts the surveys, anonymity should be preserved both in collecting and in reporting the data. When sub-segmenting the data, the number of people in a sub-segment should be large enough to prevent identifying individual respondents.

While culture surveys typically produce narrative data, it is important for statistical analysis purposes to develop clear definitions mapping narrative input to numerical scores. This will likely require identifying a range of potential responses to every question, and decide how each would fit on the scale of potential responses, perhaps from 0% to 100%.

In culture surveys, it is not unusual for some employees to respond very negatively to questions they do not really feel negative about. They may do this thinking they are punishing management, or may wish to emphasize a negative response to other somewhat related questions. In some cases, the fear of management reprisals may lead one member of the group to answer negatively on behalf of the rest, to deliver the message while saving their peers.

Therefore, a means of interpreting extreme input is also needed. This may involve observing the work group with the apparently outlying feedback or asking members of the group why some of their peers reacted as they did. Follow-up interviews may also be warranted. Of course, the negative input may also be fully accurate. Therefore, when leaders conduct follow-up interviews, they should approach it showing a sincere interest in the employees' well-being, taking negative input as an opportunity to foster trust with that part of the organization.

Schedule Assessments

The timing of a process safety culture assessment should be carefully considered to avoid unusual biases. For example, avoid asking employees to complete a survey at the end of the shift when they are tired and just want to go home. This can cause them to be resentful and answer quickly without taking time to consider the question, or just check random boxes.

Avoid assessing culture right before or after union contract negotiation periods, lay-offs, strikes, and major reorganizational changes, as results may reflect those issues rather than the state of process safety culture. Similarly, avoid periods either right before or after bonus awards and annual performance reviews. Holiday times and the summer vacation season should also be avoided because participation may be less. Finally, make sure a reasonable time has passed following a major accident, as some employees may be predisposed to negative feelings, while others may be biased to answer positively out of a form of self-defense.

Generally, the culture data collection period should be kept as short as reasonably possible. Once the process begins, employees will be keen to see the results. If the data collection period is too long, employees may begin to wonder if the effort is serious, and that, too may bias the results (Ref 6.3)

Conduct Formal Surveys

Some organizations employ formal surveys of personnel to collect information about their culture. The advantages of a written survey are that they can be planned carefully in advance with questions designed to elicit specific types of information. They leave a written record in the exact words of the persons surveyed, compared to verbal surveys where interviewers might interpret responses differently. Written surveys can also collect information from a wide group of people in less time than individual interviews, and if conducted anonymously they can help solicit more honest and accurate information. Finally, employees may hesitate to ask an interviewer to repeat questions that they could easily reread if questions are written.

If the survey questions are multiple choice, avoiding open-ended questions, the responses can be statistically analyzed, especially if they are conducted online. However, the multiple-choice format does not allow respondents to convey emotion, which is better detected via face-to-face interviews. In face-to-face interviews, interviewers can also ask follow-up questions when they sense there is more behind an interviewee's response. The best way to elicit the depth of the concern is to use a scale for responses from strongly agree to strongly disagree. Therefore, it may be valuable to use the multiple-choice on-line survey initially, and then develop focused follow-up questions to be explored through interviews with targeted groups.

Perform Individual Interviews

A meaningful culture interview requires an examination of values and behaviors. Culture interviews therefore resemble interviews performed in PSMS audits, especially in that they:

- Follow question protocols to be developed and followed,
- Benefit from developing rapport between interviewer and interviewee,
- Should include thanking the interviewee for their participation; and
- Should be documented.

However, interviews may be more challenging than audits because they involve assessing feelings rather than objective facts. Culture interviews also differ from audits in that they should involve a selection of people representing a range of job functions and positions along the hierarchy of the organization. Even simple cultural questions such as "Has the facility lost a sense of vulnerability with respect to process safety hazards?" may result in widely differing views among positions and levels (Ref 6.4).

In each interview dealing with cultural issues, the interviewer should attempt to ask questions that are purposefully indirect. For instance, the following questions might be used to probe the "sense of vulnerability" issue:

- Do you believe that a catastrophic release could happen at this plant?
- Could an incident at this plant cause damage or harm offsite?
- How would you compare the likelihood of a catastrophic release to the likelihood of a car accident? To an airplane crash? To being struck by lightning?

The interviewer may need to change the line of questioning during the interview, depending on interviewee's responses.

If a single employee's answers are inconsistent with known risks at the facility, then the interviewer may conclude that their sense of vulnerability is weak. Then by posing these questions across the organization, the interviewer can determine if, for example, workers feel vulnerable while management does not.

Interviews can be in formal settings such as conference rooms or informal such as around the lunchroom table. While interviews should usually be planned and scheduled, valuable input may also be gained from informal conversations that arise as the interviewee goes about the site.

In union facilities, hourly employees may request the presence of a union official during the interview. In this situation, the interviewer should seek to determine if the role of the union official is to put the interviewee at ease, or control their response.

While the former is helpful and may increase the value of the interview, the latter may be a clear indicator of a culture problem.

The following basic process should be helpful in establishing a framework for the overall process and increasing the effectiveness of the interviewer's on-site activities. The emphasis is placed on the interaction that develops between interviewer and interviewee rather than strictly on the mechanics of the interview process.

Plan the Interviews. The interviewer should identify the personnel to be interviewed in advance, understand the goals of the interview, determine the interview questions, and consider how to maximize the effectiveness of the interviews.

Interviews with a selection personnel that span the spectrum of responsibility will be required during a process safety culture assessment. These include representatives of:

- Senior management including the senior-most,
- Middle management,
- The process safety manager and managers of the PSMS elements
- Front line supervisors; and
- Hourly personnel including operators, maintenance personnel, and others as appropriate.

Front line supervisors and hourly personnel should be selected from each of the facility's shifts.

As much as possible, set a specific time and duration for each interview and respect the interviewee's other commitments and work schedule. Request that the facility provide coverage for operating staff in safety-critical position, and generally limit interviews with operators to 30–45 minutes to minimize disrupting operations.

Arrange a comfortable setting for interviews. Hourly personnel will generally feel more comfortable in their own working environment and may feel subtly intimidated in conference rooms usually used by management. The interview setting should be private, avoiding areas where others may be present or where passers-by may look in.

When interviewing managers, obtain a brief understanding of titles, responsibilities, and reporting relationships. This will help the interviewer understand how the process safety culture flows through the organization and where roadblocks may exist.

Individual interviews should generally be conducted by a single interviewer. This helps create a more trusting environment and avoids the potential for interviewees to feel ganged-up on by multiple interviewers. More than one interviewer could be used when interviewing executives, as they are less likely to be intimidated and it should be more time-efficient. Where hourly employees are accompanied by a union representative, talk to the representative in advance to request they provide support only and do not seek to influence the interviewee.

Design the interview protocol to provide prompts for the interviewer rather than detailed questions, and make it easy to record responses. This will allow the interviewer to focus on the interviewee rather than on the notes. Using a paper notebook or electronic tablet is generally the least intimidating to the interviewee. Since notes taken by this method will by nature be minimal, a few undisturbed minutes following each interview should be planned to record additional notes and observations.

The use of clipboards, though convenient, can convey the sense that the interviewee who is being evaluated, not the culture. Using a laptop computer as the source of the notes should also be avoided, as the screen acts as a barrier between interviewer and interviewee. Audio or video recording of the interviews should be strictly avoided.

Group interviews should include participants from the same level of the organization. This helps avoid potential reluctance to offer input in front of a supervisor. Interviewers should also be alert for dominant members of same-level groups who may dominate conversation or intimidate others. Interview groups should be small enough that all participants can offer their points of view. With group interviews, it may be convenient for interviewers to also have a note-taker to allow them to focus on the interview.

Open the Interview. The opening communication can make or break the interview. While the opening should be brief, it should place the interviewee(s) at ease. The opening should include:

- Introductions. The interviewer should begin by introducing themselves, including a brief background, and ask the interviewee to do the same. Ask the interviewee how they would like to be addressed, and use that name throughout the interview. Briefly recap the purpose and scope of the culture study and the purpose of the interview.
- Verify the timing. Ask "Is this still a good time for you?" This shows respect, but also has the practical benefit of avoiding the interviewee having something else on their mind, being interrupted, or having to leave early to attend to other business. Reschedule if necessary.
- Explain how the information will be used. Assure the interviewee that his responses will be confidential and considered only as part of a large group where individual names and responses cannot be known. Explain that the interview is intended only to understand the facility's culture and that it not a personal evaluation. Assure the interviewee that it is acceptable if they do not know the answer to the question asked.
- Request a brief overview of the interviewee's job. This serves a dual purpose. First, it informs the interviewer of the interviewee's process safety role. Second, it breaks the ice and gets them talking about something they are

familiar with. Follow-up with a more general question about how that role works.

Conduct the Interview. Work through the interview protocol, using follow-up questions to clarify answers and to assure completeness. Typically, interviewers will use three types of questions:

- *Open-ended questions* seek information in the interviewee's own words. Questions like "What does a good process safety culture man to you?" and "What needs to be done to reach your view of a good process safety culture?" allow the interviewee to provide their opinion more fully. While the answers to open-ended questions can be harder to evaluate, their information is more valuable. Open-ended questions can sometimes lead to extraneous information and tangential stories, which the interviewer can manage with other forms of questions.

- *Leading questions* help steer the direction of the conversation. A leading question like "Can you tell me more about (the desired focus of the original question) can be useful to bring a tangential question back on track. However, avoid leading question like "You follow procedures, don't you?" These can sometimes direct the desired answer or be perceived by the interviewee as a trap.

- *Closed questions* seek concrete answers, typically "Yes" or "No." These provide the most precise information but limit the respondent's ability to provide valuable detail. For example, a closed question such as "Has the Alkylation Unit PHA been revalidated yet?" may result in the answer "No." Once "No" has been stated, the interviewee may become defensive and information may be lost. However, closed questions like "Do I understand correctly what you said that ..?" can be very useful to check understanding. Take care to avoid close-ended questions that feel like a legal cross-examination.

For the most part, questions should be open-ended and neutral, avoiding wording that might influence the answer. Avoid closed, biased questions such as, "Do you want to improve process safety or maintain our current level of performance?" Questions should be worded as clearly as possible using good grammar, including terms commonly used by the company. Avoid questions that appear to have only one socially correct answer (e.g. "How often do you drive impaired by alcohol?"), as well as questions that could have two meanings. (Ref 6.3)

Interviewers should take care to maintain trust throughout the interview. While "Why" questions can sometimes be helpful in understand the cause of an interviewee's feelings, if the interviewer uses the wrong tone, the question may feel like an accusation. Likewise, the interviewer's body language should be open and receptive. A startled response to a question or an aggressive posture may put the interviewee on the defensive.

Choice of Interviewer. In general, the interviewee should perceive the interviewer as safe. The interviewer should have no real or perceived influence over the interviewee's employment. The perception of influence may extend further than realized. For example, if a neutral party conducts the interview in the office of the interviewee's supervisor, the interviewee may assume the interviewer is tied to the supervisor. Most workers view Human Resources (HR) personnel as an arm of their management chain, even if HR is in a completely different organization. Likewise, corporate process safety staff may be perceived as an arm of management even if their role is to remain independent of management. (Ref 6.3)

Maintain Respect and Trust. The interviewer should demonstrate their visible commitment to understand the interviewee's responses. They should focus on the information being shared while avoiding critical judgments about the respondent or the answers. If an interviewee provides inadequate or incomplete answers, it may reflect that they are anxious about the interview. Helping the interviewee to clarify and/or deepen his/her responses communicates respect and interest.

Probe constructively. This may be needed when interviewees provide inconsistent, conflicting, or incomplete responses. Interviewers should phrase inquiries to focus on the data rather than confronting or criticizing the respondent. If possible, the conflict should not even be mentioned. Instead draw the interviewee into the process to clarify the information.

When probing suspected negative behaviors, avoid negative and potentially accusatory questions such as, "Do you make unauthorized changes in the plant without using MOC?" Instead, pose a scenario and observe the response. To probe unauthorized changes, the interviewer could ask, "It is 2:00 AM Saturday morning. A part needs to be replaced but the replacement-in-kind part is not available. What would you do?" The verbal and non-verbal responses should reveal the true situation.

Confirm input. The interviewer should summarize or paraphrase the information learned frequently during the interview. Called active listening, it involves paraphrasing answers in the form of closed questions. Active listening clarifies the interviewee's response, while showing interest in understanding the response accurately.

Watch non-verbal signals. As the saying goes, only 10 percent conversations are verbal; the other 90 percent is tone and body language. Answers that appear inconsistent with body language or tone, and sudden changes in either may signal that the interviewer is getting close to sensitive topics (Ref 6.3)

Provide feedback, as appropriate. The interviewee may request feedback at various stages in the interview process. Because policies may vary from company to company regarding making recommendations and suggestions directly to facility personnel, interviewer should understand those policies prior to providing feedback to facility personnel. Critical judgments should be avoided.

Watch the time. Try to complete the interview on time. If extending the time would be beneficial, ask, "This is taking a bit longer than we planned, would another 10 minutes be okay?" Reschedule if necessary.

Additional techniques to help build rapport:

- *Maintain eye contact.* This connotes interest in and attention to the interviewee, and allows the interviewer to more easily read body language.
- *Maintain a comfortable distance.* Sitting close to the interviewee can make them uncomfortable while sitting too far may create emotional distance. The comfortable distance between people varies widely around the world and even within countries. Interviewers should understand this before starting interviews.
- *Mirror the interviewee.* Approximately matching the tone, tempo, and body position of the interviewee can foster rapport between the interviewer and interviewee if it is done subtly and does not look calculated.
- *Business cards.* The interviewer may present a business card as part of the introduction process. This can help establish them as independent while also conveying that they are not hiding anything. If a card is presented, it should be done casually, saying for example, "If you have any further thoughts, just call me." When interviewing senior managers, and in cultures where business cards are traditionally shared, a more formal presentation of cards may be appropriate.
- *Interviewer reactions.* Interviewers should avoid positive or negative reactions to what the interviewee has said, whether verbal or non-verbal. Positive feedback about the level of sharing may be helpful where warranted, including nodding and smiling, but all negative feedback should be

avoided. Interviewers should avoid looks of amazement or disbelief, frowns, scowls, wide-eyed looks, and startled movement. Certainly, the interviewer should not be expressionless or as still as a statue.

- *Use of silence.* In some cultures, including in the USA, there is a low tolerance for silence during conversations. However, interviewers should refrain from attempting to fill silences. The silence may arise because the interviewee is attempting to formulate their response, and if so, breaking the silence could prevent the answer. After a long silence, study the non-verbal signals of the interviewee. If they appear uncomfortable, gently ask if they need more time or clarification of the question.
- *Do not argue* with interviewees. Always be professional and courteous. If a possible finding comes up during a discussion with a management employee and immediate pushback from the interviewee occurs, politely defer a resolution until later and leave the subject to go on to the next question as easily as you can.

Close the Interview. It is particularly important to close each interview in a concise, timely, and positive manner. Thank the interviewee for their time, and for their cooperation, candor or insights, as appropriate. This will leave the interviewee with a positive impression, which they may convey to future interviewees.

Conclude the interview by asking, "Do you have any questions for me?" If there needs to be a follow-up interview this should be clear between the interviewer and the interviewer, although the time and place may not be able to be confirmed at that time.

Document Interview Results. The process of documenting interview results begins early in the interview, perhaps with a casual comment that the interviewer hopes the interviewee does not mind if some notes are taken to help the interviewer remember the information discussed. Then, immediately following the interview, take time to review the notes to ensure they accurately and completely reflect the information obtained during the interview.

Hold Focus Groups

Focus groups allow small groups of people to share opinions, thoughts, feelings, perspectives, and ideas with each other and a moderator. Group meetings can take longer than individual meetings, but can provide much more information in that time.

In focus groups, participants find comfort in numbers, and can build on each other's responses. Both can help participants express themselves more fully than they could during one-on-one interviews. Focus group meetings can also be used to develop and test the path forward for process safety culture.

Focus groups offer a window to observe at least part of the culture at work. They can reveal culture trends even without the statistical review that typically is required to analyze surveys and interviews. Focus groups may be the most effective option to uncovering a group's core values. However, focus groups typically are not large enough to represent a statistical sampling of the culture. Assessors should attempt to draw conclusions only after considering the results of a large enough number of focus groups.

Of course, focus groups are not as good at assessing individual behavior and opinions, since the group interaction will influence individuals' input. Focus group discussions can become very lively and range widely and without strong moderation can be difficult to control and document. Side topics and side conversations can consume a lot of time, and vocal individuals could dominate the discussions while others hold back (Ref 6.3).

Mind Group Dynamics. Many people behave and speak differently in a group setting than during one-on-one interviews. To help people speak more openly, assemble focus groups of people who do not know each other. This will remove the pressure for participants to not deviate from the group-think. Take care to avoid forming groups whose participants do not get along with each other. People in such groups may offer contrary opinions out of habit rather than expressing true feelings. Moderators should be alert to both possibilities and make the necessary adjustments.

Moderators need to make all participants feel safe and not pressured to answer in any specific way. They should inform participants how their comments and the overall session results will be summarized and reviewed.

Moderators need to stress that no opinions expressed are wrong, and that all participants should respect others' opinions, even if they disagree, while collecting the contrary opinions. A moderator may say, "If you have a totally different experience or opinion than the rest of the group, I need to hear it. Your view represents others who are not here today to support your view. I hope you will have the strength to speak up." The moderator should offer praise for the first contrary opinion with a comment like, "Thanks for sharing. I knew you all cannot be agreeing about this. Can we hear more?"

Plan Focus Group Sessions. The number of focus groups needed depends on the size of the site and the number of functions and levels in the site's organization. As noted above, focus groups should comprise participants of similar levels. Therefore, there will likely be one focus group of senior managers, two or more of middle managers, and increasing numbers of groups at lower levels. Each focus group should be designed around specific goals. Groups at the same level may have different goals.

Most literature recommends 6 to 10 participants per focus group, plus one moderator, plus possibly a note-taker. In larger groups, more vocal participants can drown out the input of others. The potential for side conversations also increases. Each group should include a diversity of functions, to help the moderator understand how various parts of the organization interact with each other. Smaller groups may not have that diversity, although mini-focus groups of 3 to 4 participants can be useful for evaluating individual functions.

Based on the goals of the focus groups and the structure of the site, the total number focus groups and the actual participants of each group should be identified. Schedule and location of each focus group can then be determined.

The questions for each group then should be selected. Focus groups should limit the number of questions to 5-6, to allow sufficient time to discuss each question. More than one group may, and in many cases, should be asked the same questions. If more questions need to be discussed, additional groups can be formed. Moderators should not feel pressured to get answers to all questions. Rushing at the end to answer remaining questions will not allow the depth of discussion needed. If a group fails to discuss all questions, the group should be reconvened later, or the questions asked of another group.

The focus group location should, like individual interview rooms, be familiar and comfortable to the interviewees. Ideally, the meeting room should be arranged with chairs in a circle with no table or desks in the way. This allows participants to face each other with no barriers between them. Disruptions such as radios and cell phones should be turned off or, better, not brought into the room.

In planning focus groups, realize that people who have already attended focus group sessions will talk about them. Consider encouraging them to do so. This can help jumpstart subsequent focus group discussions. Also anticipate that some attendees may linger afterwards to say things they were not comfortable sharing with the group. For the same reason, moderators should leave participants with their contact information in case they wish to communicate privately, or if they think of something else.

Moderate Focus Groups. The primary role of moderators is to get participants talking to each other. Once the conversation starts, the moderator should then ask probing questions to drive the discussion to cover the desired topics. Moderators need to quickly establish rapport with the entire group, and then maintain that rapport. If the moderator cannot establish report, any results received should be questioned. Moderators should not have a stake in the outcome of their focus groups. Their sole responsibility is to gather information and ideas from the participants.

Successful moderators:

- Treat everyone and their comments with respect and hold participants to the same standard,
- Make sure everyone participates equally and prevent any participant from dominating,
- Probe deeper into responses with phrases such as "Tell me more about that...", "I can't read the groups' reaction. Help me out", and "Boy, that got quite a rise out of everyone. What is everyone reacting to?"),
- Validate what they think they hear with phrases like, "So, it sounds like you are saying...",
- Know when to remain silent to allow others to comment; and
- Know when to encourage discussions going down a desired path.

Review records

A limited, though important, portion of the process safety culture evaluation should involve reviewing records. This can be particularly useful in detecting *Normalization of Deviance*. Records that can reveal *Normalization of Deviance* include:

- *Asset Integrity:* Are ITPM tasks being done as scheduled?
- *Management of Change:* Have Safe Work Permits been issued for changes that are not replacements-in-kind?
- *Operating Procedures:* Are exceedances of safe operating limits increasing?
- *Action Items:* Is the backlog of action items increasing? Are action items being deferred?
- *Training:* Is training falling behind or further behind schedule?

Make Observations

Observations in the facility allows process safety culture assessors to determine if information gathered through surveys, interviews, and focus groups is accurate. Assuming observation can be made inconspicuously, the true status of every process safety culture core principle can be viewed firsthand.

In leadership team meetings as well as daily production and maintenance staff meetings, assessors can see management's real imperative for process safety and if they are trying to improve process safety performance and culture. In all meetings, as well as in one-on-one communications, assessors can see if leaders are *Fostering Mutual Trust* and *Ensuring Open and Frank Communication.*

In process safety meetings, MOC reviews, and PHAs, *Sense of Vulnerability* can be readily assessed, as can *Understand and Act on Hazards and Risks.*

Observation can potentially influence behavior, especially if those being observed know why they are being observed. Which they will if the observers are the same people who conducted interviews and focus groups. If unbiased observations are required, it may be necessary to bring in new personnel to perform the observations. Additionally, the observers should be present under some other pretext.

Because observation can influence behavior, observers can also be used to reinforce key cultural objectives. This can start with known outside observers who were involved in the culture process. Then as the culture improvement effort matures and the facility gains knowledge and skills in implementing the culture core principles, employees can observe each other and offer constructive feedback.

Observations should be planned, choosing topics, specific questions, and documentation method in advance. However, assessors should have some flexibility to pursue a different line of questions if it appears fruitful.

When observing, the assessor should start by examining the messages about process safety from the employees' perspective. Messages frequently come from posters and slogans, as well as what leaders say or remain silent about. Messages also come from how leaders prioritize their goals and actions. Messages consistent with a strong process safety culture include:

- Leaders who prioritize goals and actions such that process safety is on equal footing with production and other goals,
- Posters and slogans that address process safety directly,
- Good housekeeping; and
- Corporate and site communications about process safety issues and metrics.

Messages that suggest a weaker culture include:

- Leaders whose funding priorities indicate priorities other than process safety,
- Posters that focus entirely on occupational safety,
- Poor housekeeping, stains, drips, and overall poor facility condition; and
- Communications about safety pay little attention to process safety.

Assessors should then observe operators and mechanics at work, looking for at-risk behaviors and quality of relationships and communications during workers' normal activities. Activities to observe include:

- Safe work practices: Hot work, line-breaking, equipment-opening, confined space entry, etc.
- Pre-start-up safety review meetings.
- Shift changes: If facility has separate shift changes for control room operators and field operators, observe both types.
- MOC review meetings.
- Contractor safety training (the assessment team itself might be subjected to this training to begin the assessment).
- Daily production meetings (meetings where operations and maintenance activities are discussed, scheduled, and prioritized).
- Non-routine operations.
- Safety meetings or similar events where process safety issues are on the agenda.

Once a pattern of behavior has been determined, assessors should engage in conversations with those being observed (Ref 6.3). Workers in organizations with Behavior-Based Safety (BBS) programs will be used to this for occupational safety. The goal of these discussions is to validate what was learned from surveys, interviews, and focus groups, and to identify specific opportunities to improve process safety culture or the PSMS.

Evaluate Symptoms and Causal Factors

Observations generally start by recognizing symptoms of culture gaps. From there, assessors should focus on identifying the causal or contributing factors of the symptoms recognized. Causal factors generally are determined by finding the underlying reason for the symptoms. As discussed above, avoid confrontational questions that can put the interviewee on the defensive. Once the immediate underlying reasons are known, determine the reasons those conditions exist. And so on, until the root causes are found.

Causal factor evaluation resembles the incident investigation process. While not necessary, any available root cause analysis software can be helpful in tracking causal factors and determining root causes.

Determine Reevaluation Schedule

At this writing, only Contra Costa County, CA, USA. requires culture assessments on a regular schedule. Therefore, companies may select any reevaluation timing suited to their culture improvement goals. In the early stages of a culture improvement effort, greater frequency may be beneficial, especially for the culture core principles with the largest gaps. As the organization approaches the desired culture, occasional spot checks may be sufficient.

Some companies include a high-level review of process safety culture in their regular PSMS audits. This can be a useful alert to signs that culture improvement progress has slowed or that bad habits are beginning to creep back, and a potential trigger for a more in-depth culture assessment. Remember, however, that process safety culture is fragile and can deteriorate rapidly when the organization is under stress. Therefore, leaders should consider conducting culture assessments after significant stress event. These can include leadership change, reorganization, acquisition, significant incidents, negative publicity about the site, and many more.

6.3 IMPROVING THE PROCESS SAFETY CULTURE OF THE ORGANIZATION

With the results of the culture assessment in hand, the process of improvement can begin (Ref 6.10), as illustrated in figure 6.2.

Figure 6.2 Six-Step Culture Improvement Process

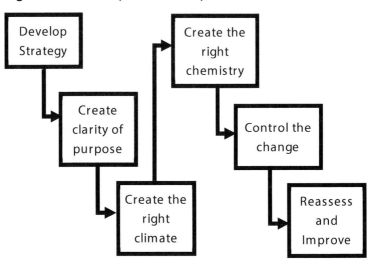

Develop a process safety culture improvement strategy.

The strategy brings the goals, aspirations, and metrics together within a defined framework. Company leaders should agree on the reasons for improving culture, create a vision and mission, and set short- and long-term goals, with metrics. Leaders should also decide how this will be marketed to the organization, including messaging, communication, and identification of influential change agents that others will follow.

If necessary, clarify roles and responsibilities for process safety within the organizational structure. Strive for clarity and simplicity, while avoiding potential conflicts of interest. Speaking about the role of unclear roles and responsibilities in the Macondo incident, Hopkins (Ref 6.11) stated, "BP has a complex structure, and BP has frequently reorganized, created new positions, and shifted responsibilities between organizations. As a result, it has not always been clear with respect to process safety just who was responsible for what."

Create clarity of purpose.

In this step, the clarified roles and responsibilities are rolled out, along with the process safety culture mission, vision, and goals. Leaders should conduct awareness orientations about the desired process safety culture for all personnel. The orientation should cover the core principles of process safety culture, the new expectations, and the plan to assess and improve culture.

The orientation should be conducted by the organizational leaders personally. Culture comes from strong, committed leadership, and a leader who is absent from the roll out, or one who introduces the orientation and then leaves may be perceived as less than committed. It is also important to involve lower-level leaders in conducting the orientation. When workers see their supervisors aligning with the new culture, they will be more motivated to follow.

Ideally, everyone should be connected in some way to addressing the cultural gaps and working towards the vision. Following the orientation, normal work teams should be engaged to address the cultural gaps identified in the assessments.

Training to improve overall process safety competency should also be done. There may be some resistance to this. Some may already believe they are fully knowledgeable, and some managers may believe they do not need to know process safety because their subordinates handle it. Leaders do not have to conduct this training (as for the orientation). However, they should show a visible commitment to the training. For example, they might personally kick-off training sessions by explaining their importance and their expectations of trainees.

Some sort of visual representation of progress should be created, and then maintained. This could be based on metrics (e.g. % of workforce involved) or on milestones (e.g. depicted on a flowchart color-coded to show items completed, in progress, not yet started, etc.) Once started, this should be continued. If the visual representation stops being updated, workers may come to think that management has lost interest – and they may be right.

Create the right safety climate.

Chapter 1 defines climate as the set of important influences in which the process safety culture exists. Climate can be driven largely through the process safety core principles of *Provide Strong Leadership*, *Foster Mutual Trust*, *Ensure Open and Frank Communications*, and *Maintain a Sense of Vulnerability*.

Providing Strong Leadership is essential. Leaders should remain steadfast in the commitment to process safety and improving process safety culture. Whenever a change is announced, employees look to see whether their leaders take the change seriously and remain committed to the change, or if it is just a flavor of the month. While strong leadership can help start the climate on the right path, leaders need to show continued commitment indefinitely to support the climate. Leaders also need to hold each other accountable for their commitment toward improving culture. This ensure consistent messages and expectations. It also deters normalization of deviance.

Fostering Mutual Trust also plays a key role. Success in process safety requires employees to be able to report problems to superiors and peers without fear of reprisals. It also requires everyone in the facility and company to gain the trust that their leaders will support them as they fulfill all their process safety responsibilities.

Likewise, *Open and Frank Communication* is required to enable all the elements of process safety to function properly. Finally, a *Sense of Vulnerability* needs to permeate the organization. This does not mean fear of process hazards. It means understanding that an incident will occur if hazard awareness is low or safeguards are not maintained.

Factors external to process safety can influence the climate, making the above activities more challenging. Poor financial performance and increasing competition may distract leaders from giving process safety as much attention as it deserves. Similarly, pending reorganization or acquisition may cause concern for workers' jobs, which takes precedence in their minds over other priorities.

Changing climate can take quite some time where negative cultures are firmly entrenched and where local societal cultures send strong messages discouraging communication or view incidents as fated. These challenges can be overcome by finding approaches that respect the local societal views. There is always a way to do this.

Ultimately, a supportive climate for process safety culture can be built based on commitment, caring, cooperation, and coaching (Ref 6.10). Good leaders do this naturally to drive other business activities. The challenge here is to apply those leadership skills to process safety as the way to manage vulnerability.

Create the right safety chemistry.

Chapter 1 describes chemistry is the "food" that nurtures the process safety culture. This food comes from the process safety core principles of *Understanding and Acting on Hazards and Risks*, *Empowering Individuals to Fulfill their Process Safety Responsibilities*, and *Deferring to Expertise*

While process safety elements in PSMSs are ordered in many ways, all rely on an understanding of hazards and risks as a basis for defining the required actions to manage them. When companies poorly understand their hazards and the safeguards needed to control them, it is hard to convince anyone of the importance of process safety.

Workers need to know that they are *Empowered* to take necessary process safety actions, particularly the actions needed to stop an unsafe process. As discussed previously, operators have a front row seat to view the process, so they are typically the first to detect a problem. They are also the closest when an incident occurs. Empowering shut-down authority and assuring there will be no reprisals for doing so goes a long way to creating good chemistry for process safety culture.

Similarly, technical and process safety experts are also well-placed to understand if processes or anticipated changes or start-ups are unsafe. *Respecting that expertise* and assuring no reprisals also is key to establish the right chemistry.

Control the change.

As the process safety culture, and with it PSMS performance improves, it is important to monitor conditions closely. In addition to the various metrics that are defined, tracking the *normalization of deviance* can serve as a useful control point. As culture improves, deviance should begin to decrease. Similarly signs of slippage in the culture can quickly be observed through increased *normalization of deviance*.

Reassess and Improve.

Chapter 7 will address the sustainability of the process safety culture.

Some of the above themes deserve additional discussion.

Leadership

Two things distinguish effective leaders: 1) the amount of time spent monitoring worker performance (work sampling) and providing appropriate feedback, and 2) listening to employees and contractors, and providing them with an environment that makes it easier for them to succeed. Generally, leaders can best achieve this with "Leadership-by-walking-around." Quite simply, leaders cannot interact with operational personnel while seated in their offices.

To be clear, this is not leadership by wandering around; it is walking around with purpose. While walking around, leaders should be thinking of enhancing and maintaining culture. They should talk with employees about process safety, provide coaching, listen to their concerns, and follow up with the needed corrective actions. Companies that practice some form of Behavioral-Based Safety (BBS) for occupational safety (Ref 6.12) may be able to simply add process safety to the BBS dialogue.

To demonstrate purpose, leaders should demonstrate that they care about the employees and about protecting them from process safety incidents. Moreover, they care about the details of process safety, ensure these details are executed properly, and resolve improvement actions in a timely manner. By doing this, over time leaders show that they are genuinely committed to process safety, while developing trust in their employees. Additionally, the more leaders do this, the more they know about how things really work, making them even more effective leaders.

Workforce Involvement

The process of assessing and improving process safety culture relies heavily on the quality of the workforce involvement. Leaders should share and discuss the results of the assessment at all levels, both workers and management. Making all levels of the organization part of the action plan to close the gaps is essential to building full commitment and ownership for the actions from top to bottom. This improves both climate and chemistry.

Turning Assessment Results into Actions

From the results of the assessments, list the improvement opportunities along with their causal factors. Then build actions to address the causal factors. This will help address root causes of cultural gaps, making the culture improvement effort more effective. If the results of the assessment identify many needed improvements or significant changes in culture and behavior, start with small steps rather than to try to overhaul the culture completely in one effort. Try to make improvements in each core principle one or two at a time. When selecting core principles to address, it may be helpful to address those at the beginning of the list first.

The Nature of Management Systems and Documentation Models

Section 4.4 discussed the institutionalization of PSMS via centralized and decentralized organizational models. Both have advantages and disadvantages. If the nature of the PSMS is found to contribute to process safety culture issues, decide whether the model needs to be more centralized or more decentralized.

The many types of documentation required by the PSMS serve a critical role in assuring PSMS performance. However, when documentation requirements are redundant, use software that is not user-friendly, or appears to not have a purpose, the stage is set for a check-the-box mentality and the *normalization of deviance*. Often, carefully designed documentation systems can make documentation easier. Involving the users of the documentation in the design process can also help in building the culture.

Communication

Communication break-downs between silos rank high among the many communication barriers discussed in section 2.4. Considering the multi-functional nature of PSMSs, connecting silos is essential to help information flow more freely between groups and individuals. This also helps reinforce the key point that the process safety culture and PSMS requires full participation and integration.

Breaking down silos can be accelerated by getting workers in one group to be interested in and familiar with the PSMS elements their co-workers in another group have responsibility for. This can lead to mutual appreciation about each other's roles and sensitivity what they can do to remove barriers to workflow between the two groups. This is particularly important for broad PSMS elements such as AI/MI and Safe Work Practices where many functions need to coordinate, and the roles between groups differ widely.

Finally, like any change, culture change can be a difficult transition for many. Increasing communication of all kinds can help people weather the transition. Regardless of their initial acceptance of the change, maintaining a steady stream of communication and encouraging inter-silo communication at least makes it clear why the change is needed and where it is going.

<u>Measurement/Metrics</u>

As mentioned above, the use of metrics to show progress of the culture improvement effort is important both to help leadership remain focused on the culture change and to show workers that progress is being made. General PSMS metrics, as discussed in section 5.1, are also critical both for operation and improvement of the PSMS. Both kinds of metrics should be discussed in leadership meetings, and communicated across the organization.

It is important to focus metrics collection on what is essential, and make the reporting, collation, and interpretation of metrics as easy as possible. Like documentation, above, metrics collection can be subject to normalization of deviance, especially if those reporting the metrics view it as a burden.

<u>Enhancing Communication</u>

As discussed previously, upward communication regarding observed issues serves a critical role in enhancing both the process safety culture and the PSMS. Leaders can do 4 things to enhance upward communication (Ref 6.12):

- *Provide indemnity*: Avoid disciplinary action related to the reported unsafe condition, as far as practical.
- *Maintain confidentiality*: Take steps to prevent identifying the reporting employee on incident reports and elsewhere.
- *Make it easy*: Remove red tape and make reporting user-friendly.
- *Acknowledge rapidly*: Thank the reporting employee and provide practical, meaningful feedback as soon as possible.

Also, be careful delivering messages. For example, in a strong culture, leaders and employees believe that all incidents are preventable. However, printing "All incidents are preventable" as a slogan on a sign may subtly convey the unintended message that management does not want to hear about incidents or near-misses. Consider messages carefully to prevent motivating the wrong behaviors.

Simplify

The more difficult employees find the PSMS, the more likely they are to seek shortcuts and *normalize deviance*. Therefore, seek to simplify the PSMS and the associated policies, practices, procedures, and activities as much as feasible. Suggestions for simplification mentioned earlier in this book include:

- *Use a risk-based approach*: Processes and units involving significant potential hazards and risks warrant a comprehensive approach. However, lower hazards and risks may be managed with a more streamlined approach. Most PSMS elements can benefit from a risk-based approach, but PHA, MOC, and MI tend to benefit the most.
- *Metrics*: As noted above, collect the minimum set of metrics and use those metrics that can be obtained easily. Where possible, automate collecting the metrics (e.g. automatically extract from operating records) and rolling up site and corporate data.

- *Emergency response*: The emergency response plan used by general personnel should be short and clear. The master emergency response plan should be organized to make it easy to find key references and resources quickly.
- *Conduct of operations*: Minimize nuisance alarms through formal alarm management. Prevent fatigue through a formal fatigue management system that considers shift rotations, overtime, and operators' personal lives.
- *Operating procedures*: Write procedures to the appropriate comprehension level, avoiding detailed explanations. Avoid large numbers of short procedures that can make it hard to find the needed procedure or keep them up to date.
- *Contractors*: Use mutual contractor screening organizations to simplify the contractor qualification process as well as recordkeeping.

Leverage Behavior Based Safety (if being practiced)

Behavior Based Safety (BBS) has been used by many companies for many years to improve occupational safety. While not intended for process safety, BBS uses some of the same techniques needed to promote a strong process safety culture.

BBS is based on peer interactions about safety. Peers observe each other's behaviors, recognizing good behaviors and offering corrections when unsafe behaviors are observed. BBS can be difficult to implement, especially when implemented in a weak safety culture. In facilities without *mutual trust,* especially in unionized facilities with poor labor relations, BBS can be viewed as an attempt by management to turn employees against each other.

However, if these challenges have been overcome and BBS is working well for occupational safety, the key features of BBS (observation, then positive reinforcement or correction of negatives) can be leveraged for process safety.

Explain the Personal Benefits

When implementing any change, nearly everyone in the organization will want to know how the change will impact them. Leaders should explain to personnel their new expectations and should help personnel understand how everyone will benefit. Short-term, measurable goals should be set, and then progress reported so everyone can have a sense of accomplishment (Ref 6.5).

6.4 SUMMARY

Improving the process safety culture of a facility starts with leaders understanding there is a problem and an improvement opportunity that it is worthy of the organization's attention. The case should be built on facts as well as on costs and benefits for improving the culture.

Once the case has been established, a baseline should be established through a culture assessment. The assessment should be built on interviews and record reviews, followed by focus groups to test improvement ideas.

The formal improvement process should start by examining the state of the culture compared to the culture core principles. The core principles should be considered roughly in the order presented, and addressed in small steps rather than trying to fix everything at once.

In considering improvements, keep things as simple as possible, and use metrics to help reinforce both the vision and progress towards it.

Above all, felt leadership – consistent and involved – needs to be sustained always.

6.5 REFERENCES

6.1 Center for Chemical Process Safety (CCPS), *The Business Case for Process Safety, 3rd ed.*, American Institute of Chemical Engineers, 2007

6.2 Baker, J.A. et al., *The Report of BP U.S. Refineries Independent Safety Review Panel*, January 2007 (Baker Panel Report).

6.3 Contra Costa County (CCC) *Industrial Safety Ordinance*, County Ordinance Chapter 450-8 (as amended).

6.4 Center for Chemical Process Safety (CCPS), *Guidelines for Auditing Process Safety Management Systems*, American Institute of Chemical Engineers, 2010.

6.5 UK HSE, *A review of safety culture and safety climate literature for the development of the safety HSE Health & Safety Executive culture inspection toolkit,* Research Report 367, 2005.

6.6 UK HSE, *Development and validation of the HMRI safety culture inspection toolkit,* Research Report 365, 2005.

6.7 UK HSE, *High Reliability Organisations – A Review of the Literature*, Research Report HR899, 2011.

6.8 Canadian National Energy Board (CNEB), Advancing Safety in the Oil and Gas Industry - Statement on Safety Culture, 2012.

6.9 Center for Chemical Process Safety (CCPS), *Vision 20/20 Self-assessment Tool*, American Institute of Chemical Engineers, 2015

6.10 Mathis, T., Galloway, S., *STEPS to Safety Culture Excellence[SM]*, Wiley, 2013.

6.11 Hopkins, A., *Disastrous Decisions – The Human and Organisational Causes of the Gulf of Mexico Blowout*, CCH Australia Limited, 2012.

6.12 Blair, E., American Society of Safety Engineers, *Building Safety Culture – Three Practical Strategies*, Professional Safety, November 2013.

7
SUSTAINING PROCESS SAFETY CULTURE

As discussed throughout this book, humans are wired to normalize deviance. This can be beneficial when deviance leads to innovation, but harmful when deviance leads to operation outside safe operating, maintenance, and technology limits. Deviance can also occur in culture. For this reason, achieving a strong process safety culture is more of a journey than a destination.

Normalization of deviance can and likely will occur, even in the high-level effort to improve process safety culture. Companies lose their *sense of vulnerability* and can tire of continuous improvement efforts (Ref 7.1), no matter how beneficial. But as much as some may desire, process safety culture cannot be treated as a project that can be checked-off as complete.

This chapter discussed ways for leadership at all levels to sustain a process safety culture improvement effort, and ultimately sustaining a strong process safety culture. So how can we sustain process safety culture?

7.1 Definition of Sustainability

Sustainability has become a popular business term with two distinct definitions:

Essential Practices for Creating, Strengthening, and Sustaining Process Safety Culture, First Edition. CCPS. ©2018 AIChE. Published 2018 by John Wiley & Sons, Inc.

1. The ability to produce profitably today without compromising the ability of future generations to do so.
2. The ability to <u>maintain</u> or support an <u>activity</u>, <u>process</u>, or results over the <u>long term</u>.

The first definition represents a mutual recognition by business and environmental advocates that society requires both profits and a clean environmental to thrive long-term. The second is used more generally in the business context and, points to strong management and leadership.

Process safety culture and PSMSs in general rely heavily on both definitions. Clearly, avoiding incidents also helps avoid environmental impacts, along with injuries to workers and the public. Also, the results of a strong process safety effort can bring additional business benefits (Ref 7.2). Equally importantly, process safety needs to be maintained and continually improved over the long term, just like other business objectives.

Some indicators of a sustainable process safety culture and PSMS include:

- The PSMS is institutionalized. It can survive the loss of its original authors, implementers, and leaders who, through their personal commitment and hard work, made it succeed initially.
- Everyone at every level is aware of their process safety roles and responsibilities, how they fit in the overall process safety effort, and how they benefit personally.
- Everyone at every level has an appropriate sense of vulnerability.
- Continual improvement of the culture and the PSMS is institutionalized and follows the Plan-Do-Check-Act model.
- Documentation is thorough enough and clear enough that any capable person can understand what has occurred in the past and plan future activities.
- Process safety culture is strong.

The last point may seem circular, but reflects a key tenet of sustainability. Sustainability must be intentional. That is why *"Learn and advance the culture"* is one of the core principles of process safety culture. More generally, each of the core principles is required for a strong process safety culture to endure.

7.2 SUSTAINABILITY OF PROCESS SAFETY CULTURE

Process safety culture, like any culture can degrade quickly without committed effort to sustain it. Almost any event, good or bad, can create conditions that unravel previous efforts. The following examples describe events that can degrade culture, how this could happen, and what leaders can do to sustain the culture.

Serious process safety incidents

Process safety incidents with severe consequences can represent a crossroads event in the life of an organization. In a strong or improving culture, leaders take the opportunity to re-examine the process safety culture and PSMS, learn and apply the lessons-learned broadly across the company, and re-commit to process safety.

However, in a weaker or degrading culture, management may turn the investigation to finger-pointing and a search for a scapegoat. In response to regulatory and public pressure, the company may seek a legal settlement. While this is a normal practice, a weaker culture will treat the settlement as evidence that the causes of the incident have been resolved. Such a settlement would not deter a stronger culture from seeking improvement.

Incidents can sometimes be caused or contributed to by an individual who takes actions that are forbidden by company policy (e.g., violating lock-out/tag-out). The investigation team should determine if it was only the one person breaking the rules, or part of a pattern where policy violations are common. If the policy violation was an unusual event, the company should not be deterred from disciplining the individual. In fact, failing to discipline would undermine the culture. The disciplinary action should be coupled with a clear explanation to the key internal and external stakeholders the reasons why such action was taken. On the other hand, if the policy violation is part of a broader trend of violations, the company should look more deeply to understand why such violations are tolerated, and correct the problem at its source.

Regulatory pressure

Facilities may find themselves receiving significant attention from regulatory authorities. This may arise in the wake of a serious process safety incident, from frequent employee complaints to regulators, or from a history of citations and fines. As a result, the facility may have frequent inspections and a lengthy list of action items.

Facilities with a strong or improving process safety culture will engage productively with the regulators to prioritize the work and manage the workload of the site to ensure that the most important improvements are made as soon as feasible. However, in weaker cultures, there may be a tendency to overload workers and seek check-the-box solutions, leading to unsatisfactory results and more *normalization of deviance*.

Changes in key personnel

Leadership is personal. When considering change in key leadership positions, a company with a strong process safety culture will focus attention on the leadership transition to avoid a loss of *strong process safety leadership*. Upon the transitions, the company will also seek to quickly establish trust in the new leader and ensure that communication channels remain open. However, in weaker cultures, the leadership transition tends to be more ad hoc. In such companies, even if the new leader is just as committed, subtle differences in leadership style may raise questions about commitment and at least temporarily weaken the culture. Recognize that new employees at all levels bring remnants of their past safety cultures with them. This requires culture discussions as part of the interview and selection processes. Other issues concerning the on-boarding process for new employees at any level are discussed later in this chapter.

Changes in the organization

Mergers, acquisitions, restructuring, and other significant organization events may have long-term positive impacts, but short-term impacts can impact process safety culture negatively. Companies with strong process safety cultures communicate frequently about the changes envisioned. This helps avoid the many distractions that can be caused by rumored and feared changes to job security, position, pay, and benefits.

Organizational changes can also lead to perceived "winners and losers," personnel whose positions improve or degrade because of the change. Companies with strong cultures recognize these individuals and manage their transitions respectfully.

Downsizing and upsizing

Companies with a strong process safety culture keep both culture and PSMS performance in the forefront when the economy slumps and later when it recovers. This helps assure that critical areas of the PSMS are resourced adequately. Companies that cut resources without detailed consideration of the PSMS or culture effectively encourage normalization of deviance. This could be seen as fatigued workers take shortcuts to complete their increased workload.

Companies that handle downsizing well sometimes fail to recognize that the same problem can occur when the economy recovers. As demand increases, the smaller staff cannot keep pace with the increased load. When new workers are hired, if they are not quickly brought up to speed on the culture and their PSMS duties, conflicts can arise that impact communication and trust.

Changes in process safety related policies and procedures:

Changes of all kinds should be expected as the culture and the PSMS improves. These inevitably lead to new and potentially unfamiliar responsibilities and activities. These can cause stress and may require adjustments to compensation and authority, along with the needed training. As culture is improved, these personnel issues should be managed carefully. New responsibilities also need to be codified in job descriptions, so they can be sustained through future personnel changes. This will help build *mutual trust* and *empowerment*. Neglecting these issues may lead to resentment, making the culture improvement effort less likely to succeed.

Lapses in leadership and failing to learn and advance the culture:

As stated at the beginning of this chapter, any lapse of leadership can lead to normalization of deviance and overall decline of process safety culture. Therefore, leadership of the process safety culture must come from the top, be encouraged by the Board of Directors, and cascade through the organization. To combat this, good checks and balances need to be in place to review adherence on a regular basis.

Central to success sustaining culture in the above examples, and in the overall life of the company, is making a firm and full commitment to continuously improving process safety culture. This commitment can be maintained through six critical success factors that are summarized in figure 7.1.

Take Cultural Snapshots

Leaders should remain alert to changes in the culture. Periodic reassessments are important, and included as one of the success factors. However, changes can and likely will occur in between formal reassessments. Recognizing minor slippage and acting quickly without waiting for formal reevaluation can head off a more precipitous decline.

Figure 7.1 Critical Success Factors for Sustaining Process Safety Culture

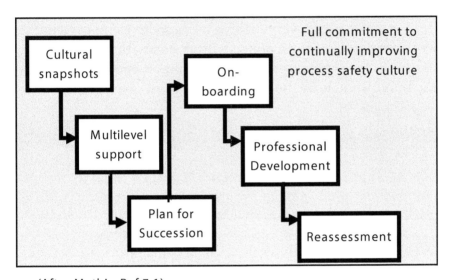

(After Mathis, Ref 7.1)

Leaders throughout the organization should seek culture snapshots in at least four ways. Snapshots can be obtained using four types of assessments, some of which may be more appropriate depending on the leaders' scope of influence. These snapshots are (Ref 7.1):

- *Focus*: Are all leaders throughout the organization on the same page regarding process safety?
- *Influence*: What factors, from job design though compensation/recognition policy, impact the ways personnel carry out their process safety responsibilities?
- *Listen*: What are people saying about process safety, and are they sincere? This may require active listening to

understand more deeply the potential impact on the process safety culture.

- *Measure*: Review process safety culture metrics and PSMS metrics and ask how these are being impacted by culture.

Assure Multi-Level Support

A sound process safety culture requires support across the entire organization in order to succeed. However, most of the problems associated with upper level support are not related to resistance but instead to multiple other competing and pressing priorities. For the culture to flourish, the support from senior leadership should be steady and consistent (See Chapter 3 on Leadership).

Also required for success is support from the lower levels of the organization, without whom the culture will be only publicity. Lack of support from the bottom is usually due to lack of understanding or lack of information, i.e., communications problems, as well as problems with mutual trust.

Plan for Succession

As discussed in Chapter 3, unplanned successions in key positions can create vulnerabilities. This goes beyond operational and process safety professionals. Indeed, senior and mid-level leadership positions are also key in this regard. Succession planning not only results in better continuity, it also tells the organization that their leaders are developing them. This creates better morale and engagement ...and better future leaders.

Onboard New Employees

Every new employee represents a vulnerability to the culture. This may be due to lack of knowledge about cultural expectations or may be from external cultural influences. The new employee orientation program should provide the core expectations for working with the culture. Then, new employees should be monitored by their supervisors to ensure they are working in accordance with these expectations.

Additionally, leaders and supervisors should be alert to new employees being indoctrinated by co-workers in ways that conflict with core cultural expectations. Awareness of this possibility should be a checkpoint for the cultural snapshots mentioned above.

Continue Professional Development

Learning to assess and advance the culture applies to individual workers as well as the company as a whole. Professional development brings in new skills and stimulates novel ideas for improving process safety. It also helps "recharge the batteries" so workers do not feel they are just going through the notions.

Professional development goes beyond receiving training. Reading, attending conferences, and making presentations are also useful. Additionally, asking employees to deliver training is an excellent way to learn even deeper what they already know.

CCPS, and other organizations provide many opportunities to attend conferences, receive training, publish and read articles, and establish peer networks.

Reassess periodically

Reassessment differs from the snapshots mentioned above in the depth of the assessment. Chapter 6 addressed the culture assessment process and how to determine the frequency of assessments. Additionally, culture status can be assessed during regular process safety audits and by investigating trends of process safety leading indicators. These may not reveal the root causes of any problems, but can trigger deeper investigation.

As of this writing, only Contra Costa County, California, USA has a statutory requirement to periodically assess process safety culture. The likelihood of other jurisdictions taking up such a requirement is not known. It is worth monitoring and learning from the Contra Costa experience.

7.3 PROCESS SAFETY CULTURE AND OPERATIONAL EXCELLENCE

In recent years, some chemical, oil, and gas companies have organized their leadership approach around the concept of Operational Excellence (OE). Each company considers a different balance of focus on process safety, EHS, business, and quality. However, every company addresses every topic at some level. Rather than treat each topic in isolation, all are addressed holistically.

Nearly all of OE efforts in the chemical, oil, and gas sectors explicitly include process safety within their scope. The following excerpts come from a few companies' descriptions of their OE programs.

Chevron (Ref 7.3)

"As a business and as a member of the world community, Chevron is committed to creating a superior value for our investors, customers, partners, host governments, local communities and our workforce. To succeed, we must deliver world-class performance exceeding the capabilities of our strongest competitors. Operational Excellence (OE) is a critical driver for business success and a key part of our enterprise execution strategy. Operational Excellence is defined as "The systematic management of process safety, personal safety and health, environment, reliability and efficiency to achieve world-class performance.

"To achieve and sustain world-class performance, we must develop strong capability in operational excellence throughout Chevron. This requires active leadership and the entire workforce to be engaged. We must develop a culture where everyone believes that all incidents are preventable and that 'zero incidents' is

achievable. With engaged and committed leadership, effective processes and an OE culture, we can achieve our objectives in operational excellence."

ExxonMobil (Ref 7.4)

"All operating organizations are required to maintain the systems and practices needed to conform to the expectations described in the OIMS (Operations Integrity Management System) Framework. To drive continuous improvement, the Framework is periodically updated. This revision strengthens Framework Expectations with respect to leadership, process safety, environmental performance, and the assessment of OIMS effectiveness and is intended to:

- *reinforce our belief that all safety, health and environmental incidents are preventable; and to*
- *promote and maintain a work environment in which each of us accepts personal responsibility for our own safety and that of our colleagues, and in which everyone actively intervenes to ensure the safety, security and wellness of others."*

DuPont (Ref 7.5)

"Operational Excellence (OE) is an integrated management system developed by DuPont that drives business productivity by applying proven practices and procedures in three 'foundation blocks' – Asset Productivity, Capital Effectiveness, and Operations Risk Management.

"The OE management system gives a company the benefits of lower costs, increased efficiencies, fewer injuries, maximum sustainable returns on operating assets, and an enhanced competitive position.

"Our integrated OE management system can be applied to existing facilities, new facilities, and facility expansions. OE gives an organization these advantages:

- *Strategic clarity about your mission, objectives, and organizational expectations;*
- *A culture of Operational Excellence;*
- *Best practices in process architecture;*
- *A well-orchestrated improvement journey; and*
- *Superior organizational alignment and execution.*

"Taking the journey toward achieving Operational Excellence typically begins with making an initial step-change improvement, followed by a continuum of incremental enhancements. Installing a culture of Operational Excellence results in significant and sustained competitive advantage."

Most OE programs are subdivided into elements, the way PSMSs typically are. Many of these elements are named similarly to PSMS elements. However, some elements go broader than process safety. For example, the Management of Change might cover environmental, occupational safety, and quality aspects of the proposed change. Other typical PSMS elements that might be combined with other topics include Incident Investigation, Training, and Documentation (e.g. Process Knowledge Management/PSI).

By doing so, OE programs find efficiencies, reducing the duplicative work that can encourage *normalization of deviance*. They also put all topics on an equal level of consideration, effectively allowing the cultures for all to be merged and not compete. OE programs also elevate these topics beyond regulatory requirements, focusing on protecting and enhancing the company rather than on checking boxes.

In other words, the cultural focus of OE matches closely the focus of process safety culture described throughout this book, while helping make the process safety culture the same as the overall company culture. OE also helps consolidate leadership focus, helping avoid the perception of conflicting priorities among all the topics.

It therefore tempting for a company to consider OE when thinking about improvements to the process safety culture and PSMS. This can make the most sense when the cultures within the various OE topics are well on their way to the desired targets. In that scenario, OE becomes the mechanism to reach the culture vision.

However, if the process safety culture, or the culture for other topics, are in the initial stages of improvement, it makes more sense to work on the topics individually to put the basics in place first. In case of any doubt, select this approach. In either case, the warning given in chapter 6 applies to OE as well: avoid trying to do everything at once, start with small, achievable goals.

The companies quoted in this section describe their OE development and implementation efforts in the context of a project led by a senior company leader or high-level steering team. The leader or team is charged with managing the deployment and transformational activities. However, once initial success is obtained, OE is no longer considered a project or program. Instead OE becomes the way the company is run, and the role of the leader or steering team transforms from project leader to leader of sustainability and continuous improvement. This is another characteristic that OE shares with process safety and other cultures in an organization: their identification as a separate and distinct program should be as temporary as possible.

7.4 SUMMARY

As stated throughout this book, the evolution of the process safety culture in an organization is a journey without end. Even if culture goals are attained, it will take leadership and continued focus to sustain it. Any thought of declaring the process safety culture effort as completed and ending work on it is an indicator of a weak culture.

On a continuing basis, plans, budgets, personnel, and leadership should be considered with the intent of sustaining culture and PSMS performance. With succession planning and continuous professional development, the organization should build a cohesive team allowing the culture to sustain even as the organizational structure and personnel change. This all takes commitment to process safety, leadership, and continual improvement, along with all the other culture core principles, which needs to be renewed from time to time.

Process safety culture needs to remain strong during economic downturns and upturns, and when a facility or company is about to be closed or sold. These are the sternest tests for the leaders of process safety culture: to keep everyone's focus on doing the right thing as they have been, despite the stresses that occur during these often-wrenching changes. Process safety culture requires real leadership.

When setting process safety goals, some companies favor priority goals, stating "Safety First" or "Nothing is More Important than Safety." Others prefer goals that are more concrete, targeting "Zero Incidents," or "X% Reduction in incidents," sometimes with a time target. Still others, despairing of ever reaching or staying at zero, favor continuous reduction goals with slogans like "Drive to Zero."

Ultimately, any of these will do. Continuous improvement to zero process safety incidents is possible. Even if there is a temporary setback in performance, it must be taken as a sign that culture efforts must be redoubled, not abandoned.

However, even if zero incidents are attained, the process safety culture journey continues. The improvement of the culture should be relentless. Good luck on your process safety culture journey, and thank you in advance for your leadership.

7.5 REFERENCES

7.1 Mathis, T., Galloway, S., *STEPS to Safety Culture Excellence*^{SM}, Wiley, 2013.

7.2 Center for Chemical Process Safety (CCPS), *The Business Case for Process Safety, 3rd ed.*, American Institute of Chemical Engineers, 2007

7.3 Chevron Corporation, *Operational Excellence Management System, An Overview of the OEMS*, (http://www.chevron.com/documents/ pdf/OEMS_Overview.pdf), 2010.

7.4 ExxonMobil Corporation, *Operations Integrity Management System*, (http://corporate.exxonmobil.com/~/media/Brochures/2009/OIM S_Framework_Brochure.pdf), 2013.

7.5 E. I. du Pont de Nemours and Company, *Delivering Operational Excellence to the Global Market - A DuPont Integrated Systems Approach*, 2005.

APPENDICES

APPENDIX A: ECHO STRATEGIES WHITE PAPER

Doing Well by Doing Good: Sustainable Financial Performance Through Global Culture Leadership and Operational Excellence

This paper may be viewed or downloaded from www.aiche.org/ccps/publications/Guidelines-culture

APPENDIX B: OTHER SAFETY & PROCESS SAFETY CULTURE FRAMEWORKS

Occupational safety culture has been the subject of many books, articles, and technical papers. However, process safety culture has received much less attention. Several publications have proposed frameworks for process safety culture along with a set of principles that define the framework. This appendix describes two of these frameworks for readers who may be interested. Both informed the ten core principles of process safety culture presented in this book. The two publications are:

- The Seven Basic Rules for the Nuclear Propulsion Program (US Navy); and
- Advancing Safety in the Oil and Gas Industry - Statement on Safety Culture (Canadian National Energy Board).

B.1 The Seven Basic Rules of the USA. Naval Nuclear Propulsion Program

This section describes the seven rules of Admiral (Adm.) Hyman G. Rickover, the founder and leader for many years of the U.S. Navy's Nuclear Propulsion Program. Starting in the late 1940s, Adm. Rickover led the effort to apply controlled nuclear fission, to the propulsion of U.S. Navy ships. Considering that nuclear fission technology stemmed from the Manhattan Project to produce the atomic bomb, Adm. Rickover and others recognized that it was potentially very hazardous.

Adm. Rickover developed a program, which had, and continues to have, a very strong reputation for strict attention to procedures and detail. The naval nuclear program represents one of the earliest examples of a high reliability organization (HRO). Appendix D contains a fuller description of HROs and their culture and the link to process safety culture.

Adm. Rickover developed his 7 Rules that serve as the core principles of the U.S. naval nuclear reactor safety. The 7 rules are

Essential Practices for Creating, Strengthening, and Sustaining Process Safety Culture, First Edition. CCPS. © 2018 AIChE. Published 2018 by John Wiley & Sons, Inc.

briefly described below, showing their correspondence to the process safety culture core principles.

In almost every way, process safety strives to accomplish the same goal as naval nuclear safety, i.e., prevent catastrophic incidents. The only difference is the added hazard of radioactivity.

- **Rule 1**. You must have a rising standard of quality over time, and well beyond what is required by any minimum standard. (*Establish an Imperative for Safety.*)
- **Rule 2**. People running complex systems should be highly capable. (*Empower Individuals to Successfully Fulfill their Safety Responsibilities, Defer to Expertise.*)
- **Rule 3**. Supervisors have to face bad news when it comes, and take problems to a level high enough to fix those problems. (*Combat the Normalization of Deviance, Provide Strong Leadership, Ensure Open and Frank Communications.*)
- **Rule 4**. You must have a healthy respect for the dangers and risks of your particular job. (*Maintain a Sense of Vulnerability.*)
- **Rule 5**. Training must be constant and rigorous. (*Empower Individuals to Successfully Fulfill their Safety Responsibilities, Defer to Expertise.*)
- **Rule 6**. All the functions of repair, quality control, and technical support must fit together. (*Understand and Act Upon Hazards/Risks.*)
- **Rule 7**. The organization and members thereof must have the ability and willingness to learn from mistakes of the past. (*Combat the Normalization of Deviance, Defer to Expertise, Foster Mutual Trust.*)

Several of these rules, particularly Rules 1, 2, 4, and 5 have direct corollaries to one of more the ten core principles or process safety culture, and all of them are related to the core principles.

B.2 Advancing Safety in the Oil and Gas Industry - Statement on Safety Culture (Canadian National Energy Board)

Following the Macondo incident in the Gulf of Mexico, the Canadian National Energy Board (CNEB) convened a committee to study process safety culture in high hazard industries including chemicals, oil, and gas. The committee studied the cultural gaps involved in Macondo and other incidents.

In the committee's paper (Ref B.1), CNEB identified several opportunities to move a concerted safety culture effort forward, including:

- Building a shared understanding of the term safety culture among regulators and regulated companies;
- Articulating clear regulatory expectations as they relate to safety culture; and
- Collaborating to develop reference and resource material for industry, to provide clarity and consistency in terminology, and describe safety culture dimensions and attributes.

The framework published by CNEB and summarized in Table B.1 describes 4 negative dimensions (i.e., cultural threats) and 4 positive dimensions (i.e., cultural defenses):

Table B.1 CNEB Cultural Threats and Defenses

Negative Dimensions (Cultural Threats)	Positive Dimensions (Cultural Defenses)
Production pressure	Committed Process Safety Leadership
Complacency	Vigilance
Normalization of Deviance	Empowerment and Accountability
Tolerance of Inadequate Systems and Resources	Resiliency

Although the CNEB framework does not map exactly to the 10 core principles described in this book, many of CNEB's observations map well to these principles. The mapping is described below.

Establish an Imperative for Safety

Attributes and descriptors of an imbalance between process safety and production, suggesting a weak *imperative for process safety* include:

- Managers make decisions based upon short-term business objectives without sufficiently considering long-term impact to process safety outcomes.
- Managers failing to see the impact of their actions in eroding process safety as an organizational value.
- Time and workload pressures.
- Excessive budgetary pressures.
- Managers less strict about adherence to procedures when work falls behind schedule.
- Project deadlines are set based upon overly optimistic assumptions.
- Frequent project overruns.
- Constant tension between production and process safety resulting in a slow and gradual degradation in safety margins.
- Shortcuts taken to meet unrealistic deadlines.
- Rewards and incentives heavily weighted towards production outcomes.

Provide Strong Leadership

Attributes and descriptors for *strong process safety leadership* include:

- Leaders participate directly in the PSMS.
- Leaders take interest in and understand hazards and risks.

- Leaders take action to address hazards and PSMS deficiencies.
- Leaders value safety efforts and expertise.
- The PSMS specifies an accountable officer with authority and control for human and financial resources.
- The PSMS specifies direct reporting lines between personnel with key process safety roles and the accountable officer.
- Timely action taken to mitigate hazards even when it is costly.
- Process safety roles receive equal status, authority, and salary to other operational roles.
- Leaders stand up for process safety even when production suffers. This usually presents an ethical dilemma for leaders. See Section 4.3 for a discussion of process safety culture and ethics.
- Safety is regularly discussed at high-level meetings, not just after an incident.

Foster Mutual Trust

Attributes and descriptors for mutual trust include:

- Everyone proactively reports errors, near-misses, and incidents.
- Policies are in place to encourage everyone to raise safety-related issues.
- Employees know and believe that they will be treated fairly if they are involved in a near-miss or incident.
- Disciplinary policies are based on an agreed distinction between acceptable and unacceptable behavior.
- Mistakes, errors, lapses are treated as an opportunity to learn rather than find fault or blame.
- Positive labor relations.

Ensure Open and Frank Communications

Attributes and descriptors for *open communication* include:

- Process safety information and performance data communicated upwards and across the organization without distortion.
- Sharing information and interpretation to create collective understanding of current status of safety and anticipated future challenges.
- Staff from a wide variety of departments and levels regularly meet to discuss process safety.
- Teams avoid making decisions in isolation; instead they seek feedback about the impact of their actions from other parts of the organization.
- A questioning attitude prevails at all levels of the organization.
- Risks and related controls are communicated throughout the organization, including contractors and, where applicable, customers.
- High quality and timely feedback is provided to staff following receipt of a report/concern.
- Incidents are thoroughly reviewed at top-level meetings.
- Lessons learned are implemented as global reforms rather than local repairs and communicated effectively to employees.
- Lessons are learned from incidents that occur across the industry and in other high hazard industries.
- Lessons learned from internal data collection are shared with others across the industry.
- Employee advocates (including health and safety committee members) have adequate training, skills, and resources.
- Employees communicate with other departments to understand safety implications of decisions.

Maintain a Sense of Vulnerability

Attributes and descriptors of complacency or weak *sense of vulnerability* include:

- Overconfidence in the PSMS and its performance.
- Inattention to critical safety data.
- Failing to learn from past events.
- Inadequate process safety data gathering, including focus on the wrong indicators and insufficient indicators.
- Performance management, incentives and rewards related to the wrong indicators (e.g. occupational injury rate) or not present at all. Control of risks weak and/or reactive.
- Supervisors do not regularly confirm that workers (including contractors) obey process safety rules.
- Organization only seeks information to confirm its superiority.
- Organization discounts information that identifies a need to improve.
- No interest in learning from other organizations or industries.
- People who raise concerns viewed negatively.
- Response to concerns focuses on explaining away the concern rather than understanding it.
- Superficial incident investigation focused on the actions of individuals.
- Failures viewed as being caused by bad people rather than system inadequacy.
- Organization believes that it is safe because it complies with regulations and standards.

On the last point, CNEB reported noted some empathy with those who felt that the process to develop regulations should lead to regulations that were sufficient to keep facilities safe. Certainly, the sufficiency of regulations can be debated. However, CNEB explained, and this book agrees, the objective of process safety culture and PSMSs is to prevent incidents. Paperwork may satisfy compliance, but if it is not completed with an appropriate *sense of vulnerability* based *on understanding and acting on hazards and risks*, it will not prevent incidents.

CNEB also noted that vigilance is the defense against complacency. Attributes of a vigilant culture include:

- Knowing what is going on through proactive surveillance.
- Understanding process safety information through analysis and interpretation.
- Everyone proactively reporting errors, near-misses, and incidents.
- Sharing information and interpretations to create collective understanding of the current state and future challenges.
- Data and metrics collection is easy and facilitated by well-functioning information systems. Only valuable data is collected.
- The organization relies on all available sources to support hazard identification. Sources include literature, vendors, communities, regulators, and more.
- Process safety information and performance data is communicated upwards and across the organization without distortion.
- Risks and related controls are communicated throughout the organization, including contractors and, where applicable, customers.
- Staff from a wide variety of departments and levels regularly meet to discuss process safety.
- Staff have non-technical knowledge and skills related to human factors, team performance and error management techniques.
- Policies encourage everyone to raise safety-related issues.

- The organization understands that a decrease in or lack of reporting does not mean that culture is strong, or performance is improving.
- High quality and timely feedback is provided to staff following receipt of a report/concern.
- Employees know and believe that they will be treated fairly if they are involved in a near-miss or incident.
- Disciplinary policies are based on an agreed distinction between acceptable and unacceptable behavior.
- Mistakes, errors, lapses are treated as an opportunity to learn rather than find fault or blame.
- Incident investigation aims to identify the failed system defenses and improve them.
- Incidents are thoroughly reviewed at top-level meetings.
- Lessons learned are implemented as global reforms rather than local repairs and communicated effectively to employees.
- Lessons are learned from incidents that occur across the industry and in other high hazard industries.
- Lessons learned from internal data collection are shared with others across the industry.
- Leadership seeks to exceed the minimum established regulatory expectations with regards to safety.
- Leadership owns process safety standards and performance and does not rely on regulatory interventions to manage operational risk.

Understand and Act Upon Hazards/Risks

Attributes and descriptors for vigilance that promote a greater *understanding and action on hazards and risks* include:

- Process safety leading and lagging metrics are collected, evaluated and acted upon. Data gathering includes third parties, such as contractors.

- Prospective analysis is conducted to identify future threats.
- Personnel are aware of the connection between cause and effect as they track the consequences of their actions and decisions.
- Teams avoid making decisions in isolation; they seek diverse feedback from other parts of the organization.
- A questioning attitude prevails at all levels of the organization.
- Leaders seek to identify and understand active failures and latent conditions that lead to accidents.
- Hazards and risks are explicitly captured, reviewed regularly, and updated.
- Risks and related controls are communicated throughout the organization, including contractors and, where applicable, customers.
- Processes are in place to ensure visibility of risk produced by a single decision/action and aggregate risk that result from multiple decisions/actions that collectively exceed safety margins.
- Leadership owns process safety standards and performance and does not rely on regulatory interventions to manage operational risk.

Empower Individuals to Successfully Fulfill their Safety Responsibilities

Attributes and descriptors of *empowerment* include:

- Employees participate in PSMS activities.
- Employees participate in setting safety standards and rules.
- Employees participate in investigating accidents and near-misses.
- Process safety is owned and communicated organization-wide.

- Everyone is willing to do what is right for process safety.
- Organizational silos are broken down.
- A candidate's process safety performance is considered in hiring, retention, and promotion decisions.
- Contractors' process safety performance is given same weight as other criteria in procurement activities.
- Positive labor relations exist.
- Employees at all levels express process safety concerns for any reason.
- Performance management systems include process safety criteria.
- Informal leaders are encouraged to promote process safety.
- Employees are held accountable and rewarded for a demonstrated commitment to process safety.
- Employee advocates (including health and safety committee members) have adequate training, skills, and resources.
- Non-operational staff recognizes their business decisions may impact process safety.
- Employees communicate with other departments to understand process safety implications of decisions.

Defer to Expertise

Attributes and descriptors for *deferral to expertise* include:

- Process safety roles receive equal status, authority, and salary to other operational roles.
- Process safety is considered at high-level meetings on a regular basis (not only after an incident).
- The organization relies on all available sources to support hazard identification. Sources include literature, vendors, communities, regulators, and more.

- Staff have non-technical knowledge and skills related to human factors, team performance and error management techniques.
- Authority to make decisions lies with the most qualified employees.
- Contingencies are in place to fill vacated roles with competent staff.

Combat the Normalization of Deviance

Attributes and descriptors indicating the presence of *normalization of deviance* include:

- The organization fails to implement or consistently apply its PSMS across the operation; regional or functional disparities exist.
- Procedures, policies, and safeguards are routinely circumvented to get the job done.
- The organization fails to provide adequate or effective systems, processes, and procedures for work being performed.
- The organization fails to provide necessary financial, human, and technical resources.
- Impracticable rules, processes and procedures, which make compliance and achievement of other organizational outcomes mutually exclusive.
- Employees find workarounds in response to operational inadequacies.
- The organization fails to provide employees with effective mechanisms to resolve operational inadequacies.
- Operational changes are implemented without management of change or PHA/HIRA.
- Rules and operational procedures are not followed.
- Extended time between reporting of process safety issues and their resolution.

- Maintenance activities not prioritized and executed as planned.
- Processes and procedures not routinely assessed for accuracy, completeness, or effectiveness.

B.3 References

B.1 Canadian National Energy Board (CNEB), Advancing Safety in The Oil and Gas Industry - Statement on Safety Culture, 2012.

APPENDIX C: AS LOW AS REASONABLY PRACTICABLE

C.1 ALARP Principle

(ALARP)For a company to manage risk consistently and apply risk management resources efficiently, it is useful to establish risk tolerance criteria and broad actions required to achieve tolerable risk (Ref C.1). Having done this, risks may be compared to the risk tolerance criteria and classified as either:

- Intolerable,
- Tolerable with appropriate safeguards, or
- Broadly acceptable.

Some companies will use different terminology or establish more classifications.

Generally, if the risk of an operation or process is found to be intolerable, it is not run until the process is changed or safeguards added to render the risk tolerable. Conversely, if the risk is found to be broadly acceptable, the operation can be carried out as is.

The ALARP (As Low as Reasonably Practicable) principle applies when risk in the tolerable zone. When the risk of a process or operation is in this zone, the facility should consider safeguards and process changes to reduce risk, and implement those measures that can reduce risk economically.

In the process safety context, the ALARP principle is implemented in the UK, stemming from the 1949 case Edwards v. National Coal Board. The regulations of the UK HSE (Ref C.2) provide guidance as to what constitutes a reasonable vs. disproportionate cost compared to the amount of risk reduction a change would provide. Some regulations in Australia also are based on the ALARP principle.

An alternative definition of ALARP, although following the same concept is contained in OHSAS 18002:

Essential Practices for Creating, Strengthening, and Sustaining Process Safety Culture, First Edition. CCPS. © 2018 AIChE. Published 2018 by John Wiley & Sons, Inc.

"An acceptable risk is a risk that has been reduced to a level that the organization is willing to assume with respect to its legal obligation, its OH&S policy, and OH&S directives." (Ref C.3)

The ALARP principle is often described using the illustration shown in Figure C.1. This is commonly referred to as the ALARP triangle.

Figure C.1 The ALARP Triangle

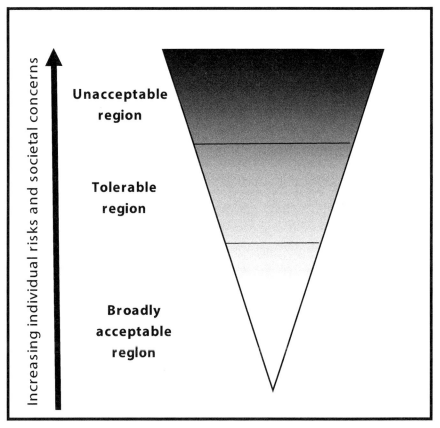

Outside the UK and Australia, the ALARP principle is not part of regulations, but is increasingly being considered as a risk management strategy, especially in companies striving to achieve a stronger process safety culture. These generally follow the concept of Safety Integrity Levels (SILs) as described in the standard IEC 61511 (Ref C.4) and national equivalents such as ANSI/ISA 84.01 (Ref C.5) in the USA.

Companies following this approach establish a risk matrix that plots order-of-magnitude probabilities against consequences of adverse events and then maps areas of intolerable risks (higher consequences and probabilities) and generally acceptable risks (lower consequences and probabilities. The remaining area of the plot are tolerable risks subject to reduction as low as feasible (e.g. ALARP), either using independent protection layers (IPLs) or process changes. Companies sometimes define timeframes in which such risk reductions should be implemented.

The ALARP concept is also used in developing risk based inspection (RBI) programs as defined in API-580/581 (Ref C.6). These standards establish a frequency of certain key inspection, testing, and preventive maintenance tasks on equipment that correspond to detecting corrosion and other damage mechanisms before they cause equipment failure and an intolerable risk.

In the U.S. the term ALARA, or "as low as reasonably achievable" has been used interchangeably with ALARP, but almost exclusively in the field of radiation protection. However, the term ALARA will occasionally be used for process safety.

C.2 References

C.1   Center for Chemical Process Safety (CCPS), *Guidelines for Developing Quantitative Safety Risk Criteria*, American Institute of Chemical Engineers, 2009

C.2   UK HSE, *Reducing Risks*, Protecting People, 2001.

C.3   International Standards Organization (ISO), *Occupational health and safety management systems - Guidelines for the implementation of OHSAS 18001*, OHSAS 18002, 2008.

C.4   International Electrotechnical Commission, *Functional Safety: Safety Instrumented Systems for the Process Industry Sector*, IEC 61511, 2003.

C.5   American National Standards Institute, *Functional Safety: Safety Instrumented Systems for the Process Industry Sector*, ANSI/ISA 84.01-2004, 2004.

C.6   American Petroleum Institute, *Risk Based Inspection*, API RP-580, 1st Ed, 2002.

APPENDIX D HIGH RELIABILITY ORGANIZATIONS

D.1 The HRO Concept

The concept of High Reliability Organizations (HROs) was introduced in Chapter 2 in the discussion of *Maintain a Sense of Vulnerability*. HROs operate nearly error-free in:

> *"...unforgiving social and political environments rich with the potential for error, where the scale of consequences precludes learning through experimentation, and where to avoid failures in the face of shifting sources of vulnerability, complex processes are used to manage complex technology. (Ref D.1)"*

Sectors associated with HROs include nuclear power, air traffic control, aircraft carrier flight operations, and hospitals.

The discussion in this section follows a literature review of 37 sources prepared by the UK HSE (Ref D.2). The review identified the following characteristic of HROs:

- The necessity to prevent a catastrophic event that could affect many people.
- Interactive complexity, i.e., the *interaction* among system components is unpredictable and/or invisible.
- Tight coupling, i.e., a high degree of *interdependence* among a system's components including people, equipment, and procedures.
- Existence of extremely *hierarchical* structures with clear roles and responsibilities.
- *Redundancy* whereby several individuals make decisions and oversee important operations. Note that the processes and systems themselves in HROs typically exhibit a high degree of redundancy, especially the safeguards and protective systems.
- High levels of *accountability* with expectations regarding strict adherence to procedures and "getting it right first

Essential Practices for Creating, Strengthening, and Sustaining Process Safety Culture, First Edition. CCPS. ©2018 AIChE. Published 2018 by John Wiley & Sons, Inc.

time" and where substandard performance is not tolerated.

• *Compressed time factors* whereby major activities may need to take place in seconds.

While the nature of these operations differs from chemicals, oil, and gas, several applicable lessons-learned about culture can be gleaned. In the above-referenced literature study, the UK HSE organized the characteristics and the lessons-learned in a figure. Figure D.1 has been modified from the original to fit in this book.

Figure D.1 High Reliability Organization Map (After Ref D.2)

Containment of Unexpected Events
• Deference to expertise
• Redundancy
• Oscillation between hierarchical and flat/decentralized structures
• Training and competence
• Procedures for unexpected events

Just culture
• Encouragement to report without fear of blame
• Individual accountability
• Ability to abandon work on safety grounds
• Open discussion of errors

Problem anticipation
• Preoccupation with failure
• Reluctance to simplify
• Sensitivity to operations

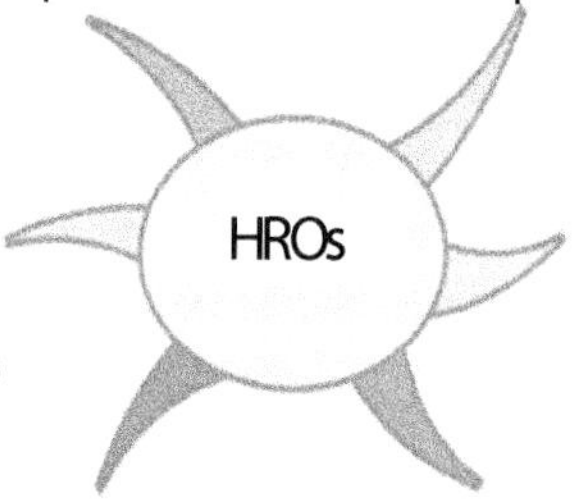

HROs

Definition
• Tight coupling
• Catastrophic consequences
• Interactive complexity

Learning Orientation
• Continuous technical training
• Open communication
• Root Cause Analysis of accidents/incidents
• Procedures reviewed in line with knowledge base

Mindful Leadership
• Bottom-up communication of bad news
• Proactive audits
• Management by exception
• Safety-production balance
• Engagement with front-line staff

HROs exhibit common characteristics that enhance their ability to deal with errors, including:

- *Redundancy*: Back-up systems, cross-checking of safety-critical decisions and the "buddy system" in which staff observe each other to catch and head-off errors.
- *Deference to expertise*: Responsibilities are clearly defined for normal operation and decision-making is hierarchical. However, in emergencies, decision-making shifts to individuals with expertise regardless of their position in the organization.
- *Empowerment*: Management-by-exception is practiced. Managers focus on strategic, tactical decisions and only get involved with operational decisions when required.
- *Preparedness*: Well-defined procedures for all possible unexpected events.

HROs effectively anticipate potential failures. HRO leaders intentionally engage with front-line staff to remain sensitive to the challenges of day-to-day operations. They remain attentive to what CCPS terms "Catastrophic incident warning signs" CCPS (Ref D.3), trivial signals that may be early indicators of emerging problems. They take warning systems, as well as performance metrics seriously, and are slow to dismiss them or explain them away.

HROs are reluctant to oversimplify. While they understand that simplicity in design is good, they also know that their operations are inherently complex. Therefore, deep understanding is required to adapt to day-to-day challenges. In such complex systems, they recognize the need to understand the systemic causes of incidents, rather than placing the blame on the operator.

HROs have a "just" culture and foster a sense of personal accountability for safety. They have systems that make it easy to report near misses and incidents without fear of punishment, and they give all employees stop-work authority. They follow-up incident investigation by implementing corrective actions.

HROs also actively seek to learn and improve. Frequent training is aimed at building deep technical competence, enabling personnel to better recognize hazards and respond to unexpected problems. Training also helps build trust and credibility among coworkers. Incident and near-miss investigations are treated as an opportunity to learn, and learnings are openly shared across the organization. Procedures are updated based on learning acquired.

HROs recognize that communications are vital, and use multiple channels to communicate safety critical information to ensure it is delivered and received, especially in emergencies. For example, nuclear powered aircraft carriers have twenty different communication devices.

HROs exhibit mindful leadership including engaging often with front line staff through site visits and active encouragement of bottom-up communication of bad news. They proactively conduct management system audits, often in response to incidents that occur in other similar industries. They also invest resources in safety management and can balance profits with safety.

Another characteristic of HROs is resilience, the ability to recover from errors. Despite their low incident rates, HROs are not error-free. Rather, they remain preoccupied with failure to better anticipate them and recover from errors when they occur.

Most of the attributes discussed above should sound familiar to readers of this book. The main differences arise from the natures of the organizations considered to be HROs compared to chemicals, oil, and gas. These differences may make some aspects of safety culture easier to attain and others more difficult.

For example, the commercial terms of nuclear power facilities are heavily regulated, with strict controls on costs, rates, and profits. In some ways, this can reduce safety vs. profitability conflicts. However, regulations are subject to politics. When regulations shrink operating margins, the conflicts could be intensified.

Additionally, the technical regulations governing nuclear power go much further in establishing design, construction, operations and maintenance standards. This is possible due to the relatively limited scope of technologies that are practiced. Again, this can be helpful in that lessons learned can be systematically incorporated into standards and communicated, but also provides less flexibility than chemical, oil, and gas facilities need.

Finally, with the unforgiving social and political environment of nuclear power, regulatory agencies place resident inspectors onsite at every nuclear power plant. This is an additional source of ever-present independent oversight with direct authority to order immediate shutdown if deemed necessary. In chemicals, oil, and gas, regulators can usually order shutdown in cases of situations deemed immediately dangerous. However, inspectors are in facilities only occasionally, and in such cases a court order may be needed.

HROs perform much more intense indoctrination of personnel than chemicals, oil, and gas. Indoctrination begins on the first day of employment, where new hires are constantly and forcefully reminded that the stakes are higher than other work places.

Training and qualification programs are much more structured in HROs than in other industrial sectors. For example, control room operators in nuclear power facilities must be granted formal reactor operator licenses based on a training and qualification process specified by regulation. In the chemical, oil, and gas sector, companies may have an internal qualification program for operators, but there are very few examples of formal training required of operators. An exception to this is a certification required by the State of California, USA, for wastewater treatment plant operators.

The limited technological scope of HROs lends itself to deep training via simulators. Nuclear plant operators, airline pilots, and air traffic control personnel can simulate and become adept at handling normal operations as well as a wide range of abnormal situations. Refresher training, via simulator, classroom, and other methods, are done more frequently than in chemicals, oil, and gas.

HROs rarely deviate from approved procedures or bypass change control processes. *Combatting the normalization of deviance* occurs constantly, including cross checking and double checking, as well as the external monitoring previously mentioned. While this may seem at odds with an environment of mutual trust, it fosters strong operational discipline, and employees in that environment consider the practice to be normal.

The naval nuclear propulsion program testifies to the success of HROs. There has never been a release of radioactive material from a naval reactor accident since naval nuclear operations began in 1955.

Early on, Adm. Rickover established a concept of reactor safety that has parallels to today's PSMSs in concept but not implementation. It consists of the 7 rules described in Appendix B, as well as 3 overarching management principles (Ref D.4):

- Technical Competence,
- Total Responsibility; and
- Facing Facts.

These three management objectives establish the cultural basis for the naval nuclear propulsion. In addition to these rules, there are 18 elements of reactor safety. These elements in effect establish the elements of the PSMS for naval nuclear reactors.

Adm. Rickover management principles have been described in his testimony to Congress (Ref D.4, and discussed further by

Paradies (Ref D.5), a process safety engineer who served in the nuclear navy. Rickover's management principles are discussed in the following paragraphs.

Technical Competence. Detailed technical knowledge of the process engineering is required not only for the design engineers but also for the senior managers, managers, supervisors, and the. Rickover said, "The more senior a manager is, the more technical knowledge (including advanced degrees and engineering qualifications) she/he must have."

Rickover insisted that senior managers understand the technical systems they managed. For example, naval commanding officers of aircraft carriers are required by law to be aviators. In addition, Rickover required that to command a nuclear carrier, officers had to receive basic nuclear power training and become qualified on the nuclear propulsion plant. This training was received along with very junior personnel they might someday command. Similar requirements do not exist in the chemical, oil, and gas sector.

Total Responsibility. Rickover lived and insisted on total responsibility. He was totally responsible for the research and development, design, operation, and maintenance of all naval reactors. This total responsibility is, in turn, passed down through the organization for each reactor, from Rickover (and his successors) to the Commanding Officer of a submarine or aircraft carrier, to the ship's Engineering (plant manager), to the Engineering Officer of the Watch (operations supervisor), to the Reactor Operator at the panel running the reactor.

Each person in the chain is totally responsible for the systems under his/her control. Anyone in the organization has total responsibility to stop operations if something goes wrong. The reactor operator does not have to ask permission to scram the reactor (initiate an emergency shutdown), in fact he/she is required to do it if conditions require.

The process safety core principles of *provide strong leadership, empower individuals to fulfill their process safety responsibilities*, and *defer to expertise* aim to establish similar total responsibility in the chemical, oil, and gas sector.

Facing Facts. This is Rickover's terminology for making difficult decisions that favor process safety and quality despite the cost, effort, delay, or potential bad press involved. Rickover said that it is human inclination to, "...hope that things will work out, despite evidence or suspicions to the contrary." He went on to say, "If conditions require it, you must face the facts and brutally make needed changes despite significant cost and schedule delays. The person in charge must personally set the example and require his subordinates to do likewise." Rickover had to "face the facts" on several well-known occasions and required extensive rework or redesign to ensure nuclear process safety.

Perhaps the most famous example of facing the facts was Adm. Rickover's decision to replace the steam piping on the USS Nautilus, the first nuclear powered submarine. Before the submarine became operational, engineers discovered that the wrong type of steel may have been used and that there was no way to verify the metallurgy of the piping in place.

Replacing the steel would cause considerable expense and delay at a time when the USA. was racing with the U.S.S.R. to be the first to adapt nuclear power to the propulsion of a submarine. Delays risked the prestige of the program and could have caused loss of important congressional support. Despite the pressure, he faced facts and insisted that the piping be replaced. Similar delays were accepted in the late 1970s on 6 nuclear submarines under construction when flaws were discovered in internal structural welding, despite immense cost and schedule pressures to commission the ships.

Weicke & Sutcliffe have summarized the success of HROs into a single characteristic called *mindfulness* (Ref D.1). Effective HROs organize socially around failure rather than success in ways that induce an ongoing state of mindfulness. Mindfulness, in turn, facilitates the discovery and correction of anomalies that could add to other anomalies and grow into a catastrophe.

In summary, traditional HROs have reached high operational discipline by convincing employees of its importance and making true believers of them. Some may have the impression that HROs do this through edict and strict enforcement. Ultimately, however, they have achieved success by changing personnel attitudes and behaviors, making process safety culture part of the DNA of these organizations.

Possibly, military (e.g., naval nuclear propulsion) and quasi-military (resident regulatory inspectors) discipline helped. While this might win minds, it does not win hearts. Winning hearts comes from a combination of factors, all of which are important:

- Carefully selecting personnel,
- Indoctrinating new hires thoroughly with a *sense of vulnerability*, including many lessons-learned,
- Training then extensively; and
- Providing continuous reinforcement from leaders who "walk the talk."

Can techniques used by HROs be applied in chemicals, oil, and gas? Maybe not all, but many can. Companies wishing to include HRO thinking in their process safety culture improvement efforts may consider the points directly above, as well as the following:

- Increasing the technical competence of management,
- Establishing total responsibility from the top of the organization to the lowest appropriate level, and holding those at each level accountable for successful and safe operation,
- Making tough decisions regarding process safety issues when required, regardless of schedule or cost impacts; and
- Training, training, training.

D.2 References

D.1   Weick, K. E., Sutcliffe, K. M., & Obstfeld, D. *Organizing for High Reliability: Processes of Collective Mindfulness*, Research in Organizational Behavior (Vol. 21, pp. 81-123). Greenwich, CT: JAI Press, Inc., 1999.

D.2   UK HSE, *High Reliability Organisations – A Review of the Literature*, Research Report HR899, 2011.

D.3   Center for Chemical Process Safety (CCPS), *Recognizing Catastrophic Incident Warning Signs in the Process Industries*, American Institute of Chemical Engineers, New York, 2012.

D.4   Rickover, H.G., Admiral, U.S. Navy, *Statement before the Subcommittee on Energy Research and Production of the Committee on Science and Technology, U.S. House of Representatives*, May 24, 1979.

D.5   Paradies, M., *Has Process Safety Management Missed the Boat?* AIChE, Process Safety Progress, Vol. 30, No. 4, 2011.

APPENDIX E: PROCESS SAFETY CULTURE CASE HISTORIES

The following case histories demonstrate the application, or failure in application, of the core principles of process safety culture. Some are taken from the public domain, some from privately-shared experiences, and some are fictional but based on real situations.

These examples are ideal for group discussion. In addition to discussing the thought-provoking questions, readers can ask "Does anything like this happen here?" and "Are there learnings we could use to improve our culture?

In nearly all the examples, the source did not analyze process safety culture impacts or did not analyze them fully. So, the actual weaknesses in cultural core principles associated with each case may not be known. Others showed strengths that should be emulated, but readers could potentially do even better.

E.1 Minimalist PSMS

A specialty chemical company produces materials using highly toxic feedstocks such as phosgene, chlorine, and several others in complex highly

> Based on Real Situations

exothermic reactions. The processes also use several flammable industrial solvents. The inventories of these feedstocks are relatively large, e.g., the chlorine is stored and fed to the process in 90-ton rail cars, of which there are always at least three onsite. The facility also changes or trials products and introduces new ones or variants of existing ones frequently.

The EHS Manager of the facility, who is responsible for the PSMS has a very long tenure and firmly believes that the best approach to complying with applicable regulations is to meet the minimum requirements and no more. He has successfully negotiated with regulatory inspectors over the years and has been successful in restricting inspections only to the specifically covered areas.

Essential Practices for Creating, Strengthening, and Sustaining Process Safety Culture, First Edition. CCPS. ©2018 AIChE. Published 2018 by John Wiley & Sons, Inc.

The PSMS does not address:

- How the significant risk of the site is managed; the focus is solely on completing the required documentation.
- Procedures and policies exist where specifically required by the regulation.
- Hazardous materials that are not covered by the regulation
- Parts of the process outside the covered boundaries that could impact risk, such as such as cooling water, power, and nitrogen.
- Equipment for processing the final products, which are not addressed in the regulations.

PHAs are performed using simple checklists because the regulations allow it, and result in little more than short memos with brief checklists attached. Audits are completed relatively quickly, and produce short reports with no more than three findings. The incident investigation file contains no investigation reports and no metrics are collected.

Do you believe the facility has had no incidents? How could they avoid them? Such an approach may reduce the regulatory exposure for a time, and it certainly may seem simpler and cheaper. However, it ignores significant risks inherent to the company's processes. The notion that strict compliance with regulations will reduce the process safety risk to a low level is a false belief and an indicator of a poor culture.

Establish an Imperative for Safety, Provide Strong Leadership, Maintain a Sense of Vulnerability, Understand and Act Upon Hazards/Risks.

E.2 – Peer Pressure to Startup

A facility has a combined MOC/PSSR process (a common practice). This process is managed electronically, routing the MOC package via e-mail

| Based on Real Situations |

to those required. However, this process calls for the PSSR, the last step in the process, to be conducted by in a face-to-face meeting. This final meeting is intended to ensure that the MOC/PSSR process does not become a "review-in-isolation," and provide at least one step of communal brainstorming regarding the change. At the end of the meeting, each participant in the MOC process must sign-off, authorizing the start-up.

The procedure specifies that the MOC Champion, who is assigned in accordance with the procedure to monitor and shepherd the MOC from its inception to its completion, and chair the meeting.

An ongoing project at the facility is several weeks late and there is increasing pressure to finish it and get the process re-started. During a PSSR meeting for this project, the Engineering representative expresses doubt about the readiness of operators to run the modified process safely, and advocated additional face-to-face training. The engineer also argues that maintenance personnel have not been fully briefed on the revised ITPM tasks will be required.

The other participants disagree, arguing that the training already provided is adequate and that the startup should not be delayed. However, this does not make the engineer feel comfortable signing-off, and the meeting is adjourned without final start-up authorization.

Later, the MOC Champion, the Project Manager, and the Operations Manager meet with the engineer's Manager to discuss the engineer's refusal to sign-off. The Project Manager states forcefully that the training requirements for were discussed and vetted by several others. He suggests that the engineer is simply being argumentative and that this is not the first time he has objected, causing delays at the last minute. The Engineering Manager agrees to sign the PSSR in lieu of the engineer.

What messages did the Engineering Manager send about process safety culture?

Defer to Expertise, Combat the Normalization of Deviance.

E.3 Taking a Minimalist Approach to Regulatory Applicability

> Based on Real Situations

A specialty chemical facility that produces many products has several product families that involve highly exothermic reactions. The facility has several normal and emergency cooling systems for the reactors that produce these products, including back-up diesel emergency generators.

The feed materials are both toxic and/or flammable and are highly volatile. The reactors process chemicals addressed by regulation, but the final products are not covered, are not highly toxic or flammable and have low vapor pressures. The reactors have multiple BPCS and SIS systems that monitor and control reactor temperature, pressure, and level, as well as dual relief devices.

The facility defined the regulatory boundaries of the facility to include all equipment from raw material storage to just before the first valve downstream of the reactors. They argued that since the products were not regulated, the equipment handling them need not be addressed in the PSMS. Note that the valve is a remotely operated by instrument air and opens and closes automatically based on the temperature in the reactor.

The facility also excluded the cooling systems for the reactors, including the backup power systems from the PSMS, since water and power are not regulated, and in any case, other reactor safeguards protect the reactor in case of thermal runaway. The regulatory manager corporate legal have reviewed and approved the PSMS boundaries.

What could be the impact of excluding utilities, back-up power, and the downstream valve from the PSMS?

Establish an Imperative for Safety, Provide Strong Leadership, Maintain a Sense of Vulnerability, Understand and Act Upon Hazards/Risks.

E.4 Not Taking a Minimalist Approach to Process Safety Applicability

| Actual |
| Case |
| History |

A specialty chemical facility that uses a dozen different flammable solvents stores them in small, low pressure storage tanks that are kept at a slight overpressure by the presence of a nitrogen blanket. Nitrogen is supplied by a vendor-owned system that vaporizes liquid nitrogen and supplies it to the facility's distribution system where the pressure is reduced and regulated for various uses.

The facility PSMS Manager included the liquid nitrogen supply system in the PSMS because the loss of nitrogen blanketing could allow air to enter the vapor spaces of the tanks, possibly creating an ignitable atmosphere. This decision was made even though the regulation would have allowed exempting these tanks and thee nitrogen supply from the PSMS.

The PSMS Manager also worked with the vendor to determine which elements will be the facility's responsibility and which will be handled by the nitrogen vendor. This, too, was not required by regulation.

This example clearly illustrates *Establishing an Imperative for Safety*. What other culture core principles are illustrated?

E.5 What Gets Measured Can Get Corrupted

| Based on |
| Real |
| Situations |

As the old saying goes, "What gets measured gets managed." Today, reporting and analyzing key performance indicators (KPI) have become a normal business activity. Increasingly, metrics extend to EHS management and PSMS. However, KPI metrics are not just dispassionate data. Collecting, analyzing, and acting on them are very human activities and can be fraught with cultural concerns.

A facility included a KPI based on the number of overdue ITPM tasks in the AI/MI element, which many facilities do. The facility defined the KPI as any ITPM task that was overdue in 2 main asset-tracking software packages. One software was used to manage rotating equipment, instruments, and electrical equipment, while the other was used to manage fixed equipment pressure vessels, tanks, piping, and relief devices. Upon implementation, this KPI revealed a few items overdue month-to-month, but the value was low, as was the aging of the overdue ITPM tasks.

Two years later, during a PSMS audit, auditors found that there were other ITPM tasks that were important to process safety that were overdue but were not tracked in either of the two tracking software packages and therefore were excluded from the KPI. And those results were much less favorable.

The Fire Chief tracked fire system ITPM in his electronic calendar. The annual fire pump flow tests had not been conducted for two years and the ITPM tasks required by NFPA-25 were not included in the calendar. The Instrument shop supervisor tracked the annual calibration of testing equipment in a spreadsheet and there were ten pieces of test equipment that were overdue for annual calibrations.

Separate ITPM monitoring systems were also maintained for vibration monitoring, electric power distribution equipment, and equipment required for the emergency response plan, and in all systems, many important ITPM tasks tracked by this system were found to be either overdue, missing from the system, or both.

The Plant Manager was surprised and upset when these findings were presented at the audit's daily debriefing. When the ITPM KPI was updated to include all the missing data, the performance was much poorer. More importantly, much work and expense were needed to catch up.

Failing to include the data from the other sources was found to be an innocent mistake. However, why was the definition of the KPI not reviewed for completeness? Why were positive results not challenged to ensure they reflected reality?

Combat the Normalization of Deviance, Understand and Act Upon Hazards/Risks.

E.6 KPIs That Always Satisfy

A facility tracks an overdue ITPM metric monthly. The data is reported to a corporate process safety metrics program, and the KPI is analyzed and published for everyone in the company to see. The values for all facilities, since the metrics program was established three years ago have been consistently above 99% completed on time, which the company was proud about result.

Based on Real Situations

The facility had just undergone a major turnaround that had been planned to be 3 weeks but had been shortened by 5 days due to production pressures. The month following the end of the turnaround, the overdue ITPM KPI still showed 99.6 % ITPM completion. Upon closer review it was discovered that 75 ITPM tasks scheduled for the turnaround had not been performed due to the shorter time. This included many proof tests of SIS and BPCS functions.

The overdue ITPM KPI did not reflect these unperformed tasks because they had been reclassified in the maintenance management system as turnaround maintenance tasks and not recurring maintenance tasks, while the KPI only considered recurring maintenance tasks.

Planned turnarounds do get shortened. However, some ITPM tasks can only be performed during turnarounds. What can be concluded about a facility that does not consider turnaround ITPM tasks in its ITPM KPI? Do you think ITPM was considered in the decision to shorten the turnaround? If business considerations really required shortening the turnaround, what should the facility have done to ensure that turnaround ITPM was conducted?

Combat the Normalization of Deviance, Understand and Act Upon Hazards/Risks.

E.7 Abusing ITPM Extensions/Deferrals

A facility tracks an overdue ITPM metrics monthly. The values for the facility since the metrics program was established three years ago have been consistently above 99% completed on time. During an internal audit, the PSMS Coordinator discovered that dozens of fixed equipment ITPM tasks had not been performed. These included external inspections of pressure vessels, internal and external inspections of storage tanks, external inspections and thickness measurements piping, and relief valve maintenance tasks had not been performed. However, the missed tasks were omitted from the monthly KPI for overdue ITPM.

The reason is that the facility had an ITPM extension procedure that allowed ITPM tasks that were due to be deferred to a later date under certain conditions upon approval of the Maintenance Manager. The PSMS Coordinator found dozens of open extensions, some of which had been place for over a year. These extensions were excluded from the ITPM KPI data.

When this became known, the facility added another KPI based on the number of ITPM tasks with open extensions and their aging and a different picture emerged. With further study, the Process Safety Coordinator also found that over the past 5 years approximately 50 relief valves whose maintenance has been extended had failed their pop test.

The Process Safety Coordinator then reviewed the extensions associated with safety instrumented systems and discovered that the 65 proof tests for SISs had been extended over a 5-year period, including 6 proof tests that were currently overdue. The overdue proof tests of SISs voided the SIL calculations for the SISs involved resulting in a higher than allowable risk to exist.

These discoveries caused the facility to review their policy for extending ITPM and to provide limits of the extension periods. The facility also excluded the ITPM of certain types of the equipment from being deferred without Plant Manager approval.

What should the Plant Manager should have done to address the abuse of the extension/deferral policy?

Combat the Normalization of Deviance, Understand and Act Upon Hazards/Risks.

E.8 The VPP Defense

The Voluntary Protection Program (VPP) is a program of the USA. OSHA. Facilities that cooperate with OSHA and meet certain proactive safety management criteria become recognized as "VPP Star."

Actual Case History

A facility in the USA has several regulated processes. While auditing the Asset Integrity element, an auditor discovered that the internal inspections and wall thickness measurements for 5 pressure vessels out of two dozen are overdue, in some cases by a few years. The same recurring maintenance tasks for 3 of 12 low-pressure storage tanks are also overdue, again by a few years.

The Maintenance Manager explained to the auditor that the plant is considered a safety model regionally, it has never been cited for overdue vessel and tank inspections. He also states that the site has been an OSHA VPP Star site for nearly 10 years, and the relationship with the local OSHA field office is excellent.

The time and effort to quickly perform the overdue vessel and tank inspections will be substantial and will result in some unscheduled down time and late product shipments. The Maintenance Manager and Plant Manager are firmly opposed to incurring these production outages on what they believe to already be a "best in class" operation.

How relevant is special recognition such as VPP Star, OHSAS 18001 certification, etc. to the extent to which a facility is managing its risks adequately? How can you segregate recognition that can boost a company's image to the public from KPIs that more accurately define the process safety performance of the facility? Is there any recognition that can serve as a "free-pass" for operational discipline?

Combat the Normalization of Deviance, Maintain a Sense of Vulnerability, Establish an Imperative for Safety.

E.9 Double Jeopardy

A new Process Safety coordinator attends a PHA revalidation shortly after she started working at a facility. It is being led by a process engineers who has worked at the facility a long time and led many of the facility's PHAs. She discovers that many hazard scenarios she believed should be included were discarded by the team because multiple failures would have to occur to realize the scenario.

| Actual |
| Case |
| History |

The team leader and the rest of the team seem to resist including these scenarios. In a later discussion with the team leader, she learns that this is way PHAs have been performed at the facility for years, and none of the many auditors and government inspectors had challenged the "Double jeopardy" assumptions before.

The Coordinator explained that at her previous facility, multiple failures were considered possible, and were considered in PHAs. The team leader seemed to regard this difference of opinion as a minor technical detail while the Coordinator regarded it was a fundamental flaw.

The new Coordinator was almost certainly right. Many deviation - consequence scenarios identified in a PHA require 2 or more Independent Protection Layer (IPLs) to reduce the risk to a tolerable level. Indeed, many incidents involve multiple failures. The Bhopal incident involved up to seven failures, although some of the failed layers of protection were not independent.

What cultural conditions make the concept of double jeopardy attractive? How should the Coordinator convince her colleagues to evaluate PHA scenarios more thoroughly?

Understand and Act Upon Hazards/Risks, Combat the Normalization of Deviance.

E.10 Best Case Consequences

During the same PHA described in example 9, the new Process Safety Coordinator also discovered that the consequences for some

> Actual
> Case
> History

scenarios were not the credible worst-case consequences. The consequences recorded in the study assumed that some of the safeguards limited the severity of the consequences.

The team leader explained later that the company considered these safeguards highly reliable. They had never failed and were regularly tested and inspected. The team leader believed that it would not be reasonable to discount them. Indeed, he felt insulted that someone who knew nothing about the facility's PHA approach would challenge his previous PHAs.

The new Coordinator eventually won the PHA leader over to her point of view, and the facility approach was changed. What techniques do you think she used to convince him?

Understand and Act Upon Hazards/Risks, Establish an Imperative for Safety.

E.11 New Kid in Town

Based on Real Situations

A new process safety engineer was performing an audit of the MI program of a facility. The Maintenance Manager had 35 years of experience at the plant and the Chief Inspector had 25 years of experience. The engineer found a very small number of overdue ITPM tasks that had not aged very long.

In a meeting to discuss his findings, the Maintenance Manager and the Chief Inspector called the number overdue trivial, and said the engineer was being overly picky. They reminded the young safety engineer that they have been around many years and this was the best ITPM completion performance they ever had. Certainly, these findings should not be in the audit report.

Should they be? What symptoms of culture problems does this scenario exhibit?

Combat the Normalization of Deviance, Establish the Imperative for Process Safety

E.12 The Blame Game

Based on Real Situations

An incident investigation resulted in two facility personnel being suspended without pay two weeks. A year later another investigation also resulted in disciplinary action. Both incidents involved fires that caused significant property damage, lost production, and some minor injuries to facility personnel. The local fire department responded to both incidents, and the media coverage was strongly negative.

The Facility Manager stated at a safety meeting that the disciplinary actions were the only way to instill a firm sense of accountability. Following these investigations, the level of participation and cooperation in any safety related activity decreased markedly, especially in incident investigations, including near misses.

Why did the participation and cooperation in safety efforts, especially incident investigations, drop off? Do you think that the facility found and corrected root causes of incidents?

Ensure Open and Frank Communications, Foster Mutual Trust.

E.13 Conflicts of Interest

Based on Real Situations

The PSMS Coordinator reported to the Operations Manager. This structure was intended to reinforce the message that all process safety matters were line management responsibilities. In this structure, the PSMS Coordinator served as an advisor to the Operations Manager.

However, many decisions advocated by the Coordinator were subordinated to production concerns. On several occasions when the Coordinator attempted to include process safety concerns into the decision-making process, he was asked to prove that an unsafe situation existed.

In one case, the Manager decided to skip a scheduled short outage to proof test safety instrumented systems (SIS), among other process safety activities. The Coordinator explained that these were SIL2 control loops that required a specific proof testing frequency to provide the expected amount of risk reduction. Failing to test would introduce an intolerable risk because the SISs could not be assumed to have the required level of reliability. Few others in the room understood the technical issues associated with deferring the proof testing of SISs. The concerns of the PSMS Coordinator were summarily dismissed.

It appeared that the facility was on the right track by having the PSMS Coordinator report to the Operations Manager. Where did they go wrong?

Establish an Imperative for Safety, Defer to Expertise.

E.14 No Incidents? Not Always Good News

Based on Real Situations

The monthly KPIs for process safety incidents and near misses at a refinery had been very low for several years. The new Refinery Manager was pleased with this KPI, especially since in his first year it was zero. In his previous refinery where he had been the Operations Manager, the same KPI had been favorable but not that good.

He asked the PSMS Coordinator how the KPI was derived. He learned that during acquisition negotiations five years earlier, the previous owner had been challenged by several potential buyers about the high rate of near misses. The near misses were not serious and no actual incidents had occurred, but the company attempted to lower their bid because of it.

After the acquisition, the refinery began investigating and addressing near misses less formally. Consequently, when the KPI program was put implemented, the near miss result was very positive.

Further review revealed that during the previous two years several SISs had been activated during plant upsets or transients. These had not been classified as near misses because, according to an e-mail, "the safeguards had worked as designed and that's not a near miss because that was what they are supposed to do."

Following this discovery, the facility revised the definition of the near miss KPI to align with the API and OGP standard for near miss reporting. This standard recognizes that a SIS trip usually represents a close approach to the capability of the equipment to contain the process, and therefore truly a near miss. By tracking these types of near misses, the facility has an opportunity to learn about the process, culture, and PSMS without suffering any adverse consequences. As a result, the data reported monthly returned to values that were more typical for a large refinery.

This example shows both good and bad examples of the role of leadership in process safety culture. What are they?

Combat the Normalization of Deviance, Provide Strong Leadership, Maintain a Sense of Vulnerability.

E.15 Check-the-Box Process Safety Management Systems

Based on Real Situations

A corporate process safety audit found that the documentation for key process safety activities at a facility was extremely sparse. Previous internal audit reports consisted of 2-page memos. PHA reports of major process units contained 10 pages of worksheets and these contained many blanks. Incident investigation reports contained root cause analyses that were described in a brief paragraph.

Further interviews revealed that these documents were created as the result of activities intended mainly to get activity off the facility's to-do list. The auditors pointed out that such practices and the thin documentation did not reflect typical industry practices for those PSMS elements.

The Facility Manager and members of his management team reacted angrily. They stated forcefully that the facility had never suffered a process safety incident and that their documentation met the minimum regulatory. This, they said, was proof enough that no additional effort was required or needed.

What other symptoms of weak process safety culture do you believe existed at this facility?

Combat the Normalization of Deviance, Understand and Act Upon Hazards/Risks, Establish and Imperative for Safety, Provide Strong Leadership.

E.16 There's No Energy for That Here

Actual Case History

During PHAs at a facility, team leaders typically screened the recommendations made by the team for their potential acceptance/reaction by management. They deleted any recommendation thought to be too expensive, time-consuming, or difficult. Occasionally, the risk rankings were re-assigned so that recommendations not be necessary.

During an audit, interviews with some of the team leaders revealed that they believed that it was their responsibility to make the recommendations addressing problems identified in the PHA go away. When pressed further about why not make the problems go away by truly addressing them, each responded "There's no energy for that here." The team leaders believed management did not want to be the ones to decide not to address a recommendation. Some believed that their performance would be adversely evaluated if they submitted PHA reports with major recommendations.

In several cases, PHA's were re-convened to revise the risk rankings and recommendations to make them less onerous or unnecessary.

Who has the responsibility to choose between implementing recommendations or accepting risk?

Establish an Imperative for Safety, Understand and Act Upon Hazards/Risks, Provide Strong Leadership.

E.17 Not Invented Here

A new PSMS Coordinator attempted to incorporate several good practices from the facility where he previously worked. He believed the facility could

> Based on Real Situations

benefit from these ideas and that they would be a relatively good fit with his new site's PSMS, personnel, and policies.

His manager disagreed, saying that the Coordinator's previous company was different, the practices were actually poor fits, and they would be too time-consuming and upsetting to implement something different when the current PSMS seemed to be running smoothly.

Over time the Coordinator noticed that the PSMS elements had become rigid and that the Manager resisted any improvement ideas, regardless of the source. The Manager even rejected improvement suggestions from the corporate process safety team. It became clear that the Manager's inflexibility was simply protecting his turf.

When is being rigid about maintaining consistent practices good, and when it is bad?

Provide Strong Leadership, Empower Individuals to Successfully Fulfill their Safety Responsibilities, Defer to Expertise.

E.18 PHA Silos

A large facility performed complete PHAs that comprehensively identified and identified controls for process safety risks. These studies were

> Actual
> Case
> History

carefully revalidated over the years to keep them up-to-date. The recommendations were resolved promptly and there were good records of these practices.

However, a closer look at PHA practices revealed that although recommendation management was excellent, the thoroughly performed PHAs were not used for any other purpose. The AI/MI team did not receive the report so they could ensure that critical equipment identified in the PHA was included in the MI ITPM and QA programs. The training team was unaware of why recommended training was needed. The emergency response planning team was unaware of the potential consequences they needed to plan for. The PHA program, as good as it was, had become a silo activity.

How can this happen in a large facility? How can cross-fertilization be encouraged when the PHA team is in a completely different organizational structure and where so many people are involved?

Understand and Act Upon Hazards/Risks, Provide Strong Leadership, Ensure Open and Frank Communications.

E.19 Knowing What You Don't Know

A new Facility Manager came from a business background. She had no experience or training in engineering or operations, and little working

| Based on |
| Real |
| Situations |

knowledge of process safety technology or management systems. Her facility had a procedure that required the Facility Manager to sign permits approving bypass of critical safeguards, including SISs and relief devices. The Facility Manager was also required to approve extension of ITPM tasks for the same types of equipment.

Shortly after taking the job, she received several requests to extend the proof testing of a SIS by 6 months and to bypass a relief device by shutting the inlet and discharge block valves. She did not know what a SIS was, only vaguely understood pressure relief, and was unfamiliar with the process safety ramifications

The requests were presented at the start of a long operations meeting with a very full agenda. It was clear that they represented critical maintenance tasks that were delayed pending her approval. With a full agenda ahead, she signed the permits, even though she did not understand the risks involved. She justified signing to herself thinking that the requesters would not ask if they did not think it was safe.

Non-technical managers do get assigned to senior operations roles. What preparation should they have before assuming those roles? What are some questions the new operations manager could have asked to be more informed when signing the permits? Regardless of background, a new facility manager cannot be expected to know everything about the facility. How can facility managers and their teams bridge this knowledge gap?

Understand and Act Upon Hazards/Risks, Defer to Expertise, Provide Strong Leadership.

E.20 Bad News is Bad

A Facility Manager prided himself on running a very tight ship. He had a bad temper and did not react calmly to negative events or people who disagreed with him. His direct reports dreaded the daily operations meetings because even minor problems led to harsh interrogation.

Although bad news cannot remain hidden for long, the Manager's direct reports went out of their way to avoid bringing bad news to these meetings. Instead, they attempted to solve problems offline, usually alone to avoid having anyone who might leak the news from being aware. Root cause analyses of several incidents and near misses uncovered this lack of broad discussion as a contributing factor.

What recommendations could the incident investigation team make to address this contributing factor?

Provide Strong Leadership. Foster Mutual Trust

E.21 The Co-Employment Trap

The legal concept of co-employment was developed to prevent long-term contractors who act essentially as employees from being denied the same benefits available to employees. While a complex concept, co-employment occurs when contractors are treated the same as employees, except in the benefits available to them.

To steer wide of co-employment concerns in a facility, contractors, including even resident contractors, were excluded from all employee activities. Contractors could not attend daily production meetings, toolbox meetings, training or safety meetings.

The facility used many resident contractors to supplement facility personnel. These contractors worked at the facility every day, in the same group as employees, and doing similar jobs. The only difference was that their paychecks and benefits came from their own employer, not the host facility's company.

Because of this strict policy, the resident contractors in the instrument shop did not attend the daily toolbox meeting and did not receive some key process safety related information.

Consequently, a resident contractor instrument technician made an error performing a proof test, and a minor incident resulted. The root cause analysis revealed that facility instrument technicians received the specific knowledge that was given to the at the toolbox meeting, but contract instrument technicians did not. The company expected that this information would be relayed through the contractors' employer, but it was not.

How can leaders work with the Human Resources function to assure contractors receive needed process safety information?

Ensure Open and Frank Communications.

E.22 Stop Work Authority/Initiating an Emergency Shutdown

> Based on Real Situations

A high-risk facility has clearly written procedures giving on-duty operators the authority to initiate an emergency shutdown when the conditions warrant, without obtaining any other approval. A review of operator training materials shows that this authority is clearly and explicitly stated. Operators have stated in interviews that the procedures were known and understood and confirmed that the training was given as it appears.

Despite this policy, an operator on duty in the control room did not initiate an emergency shutdown during a significant transient event in which the process pressure and temperature rose rapidly due to a runaway reaction. This incident resulted in a significant release of flammable materials and a vapor cloud explosion that resulted in injuries to personnel and significant property damage to the facility.

In the investigation that followed, the operator stated that he did not feel comfortable taking SWA action and that a supervisor should have been there to make that call. When asked why he was not comfortable, the operator responded that in over the years, when SWA was used, there had been a lot of second-guessing by investigators after the fact.

Further review showed that the incident investigation reports described alternative actions that the operators could have taken in response to the indications they were receiving at the control board that would have abated the transient but kept the process running. Some reports also suggested disciplinary action, although none was taken.

When operators exercise SWA, it is certainly possible that options existed for them to bring the process under control. But under duress, it is hard to know if such an option exists or not, which is why SWA is so important. How can incident investigators address potential alternative actions without undermining SWA?

Foster Mutual Trust, Combat the Normalization of Deviance.

E.23 SWPs by the Numbers

Safe work permits are involved activities used to help ensure that the hazardous work is fully prepared before any work begins. In a large facility,

> Actual Case History

these permits (e.g. Safe Work, General Work, Hot Work, Confined Space Entry, Line Breaking, and others) had been issued by the on-duty operators. The very large number of permits being sought at the beginning of day shift would overwhelm the board operator and completely distract him from running the equipment. To address this problem, the company appointed a set of permit approvers especially for this "rush hour."

These approvers were three other Operations personnel, who sat in a conference room in the control room building. Workers seeking permits lined up to see one of these three approvers. The approvers reviewed the permit presented to them and then signed. This allowed them to work very efficiently, allowing day shift work to begin.

However, the three permit approvers never left the conference room during this process. They cannot from there observe the physical location of the work, and in their rush to issue permits they ask only cursory questions of the permittees.

What cultural gap allowed the SWP process to go from bad to worse?

Combat the Normalization of Deviance, Defer to Expertise.

E.24 Incomplete MOC

Actual
Case
History

A specialty chemical facility had a several reactors that make various products. When the research group needs the trial a new product or when a new product is launched, an MOC was written addressing the introduction of the new products.

The MOC considered the new feed materials and processing conditions are examined. However, it did not normally require that capability of the rupture disks be verified to ensure they were adequately relieve the pressure excursions that could occur with the new chemistry. This would only happen when a PHA was performed and a high-pressure deviation was identified.

What holes exist in this approach?

Combat the Normalization of Deviance, Understand and Act Upon Hazards/Risks.

E.25 Post-MOCs

Actual Case History

A large facility with multiple units processed approximately 100 MOCs each month. The PSMS Manager for the facility ran the MOC program in addition to being directly responsible for several PSMS elements and being deeply involved in the remaining elements.

An audit revealed that a several MOCs had been approved after the physical change had been made. During interviews, with the PSMS Manager and others were not much concerned about this and it apparently had been the norm for years. The prevailing belief was that MOC was satisfactory if the documentation was complete.

How can a facility cope with a large flow of MOCs and still treat each one with the appropriate *sense of vulnerability?*

E.26 Mergers & Acquisitions

Actual Case History

A large chemical facility was in the process of being sold to a competitor. The acquiring company was in the process of a due diligence review of the organization's operations, including a thorough review of the status of EHS programs. The acquisition was being closely monitored by the local community, labor unions, political leaders, and the media because of the long history of operations by the facility and the many jobs that were at stake if the acquiring company decided to withdraw from the deal.

A regular audit that had been scheduled came due just as the negotiations and due diligence process began. There were recommendations to postpone the audit but there were regulatory implications of doing that so the audit was conducted as scheduled. The PSMS was found to be in fairly good shape, but the auditors did discover a few important findings.

One PHA revalidation was several months overdue, several PHA and incident investigation recommendations were somewhat overdue. Three corrective actions from the previous audit had not been completed. In total there were 35 findings, which was not an unusually high number based on previous audits performed by the same corporate staff.

However, the atmosphere for the audit was very tense, with significant pushback at each daily debrief. These debriefs were attended in person or via telecon by facility management, as well as senior corporate managers, and corporate legal staff. The audit team leader was pressured to reduce the number of findings or write them in a way that minimized their impact.

Extensive debate with sometimes heated discussion occurred challenging the interpretation of the regulatory and corporate requirements. Some of this re-interpretation involved issues that presumably had been settled practices within the company for a long time. The team leader was able to delay any final decisions on the number and nature of the findings until the closing meeting. At the closing meeting, the pressure continued.

The audit leader refused to relent. After the closing meeting, the audit leader received a call from his boss telling him that he would take responsibility for the contents of the draft and final audit reports.

What do you think of the conduct of the audit leader? What could he have done after his boss took over drafting the audit report?

Combat the Normalization of Deviance, Understand and Act Upon Hazards/Risks.

E.27 Poor Understanding of Hazard/Risk Leads to an Even Worse Normalization of Deviance

| Actual |
| Case |
| History |

An explosion and fire at a polyvinyl chloride (PVC) manufacturing facility killed five workers and severely injured 3 (Ref E.1). The explosion and fire destroyed most of the reactor facility and adjacent warehouse. Smoke from the fire drifted over the local community, and as a precaution, local authorities ordered community evacuation lasting two days. It was the third incident with a similar cause experienced by the company.

The incident investigation found that while cleaning an out-of-service reactor, an operator forced open the bottom valve of the wrong reactor, bypassing a critical safety interlock by attaching an air hose adapted to fit an instrument air connection to the "open" port of the valve. A label on the hose described it as an "Emergency Air" hose. The contents of this reactor, hot reacting vinyl chloride monomer and partially formed PVC, drained onto the floor. Shortly afterwards, the flammable mixture ignited. The resulting fire killed the operator, his supervisor, and 3 other operators.

The investigation found that the "Emergency Air" line was provided to allow operators to drain the reactor in a runaway reaction scenario in case the normal vent and relief system alone were not sufficient to control the pressure during a runaway reaction. It seems clear in hindsight that the drained mixture would have ignited as occurred in this incident and therefore may not have provided much mitigation benefit. Instead, the "Emergency Air" line had become routinely used for what the operator thought he was doing – forcing open the bottom valve of a reactor being cleaned, rather than opening it according to procedure, from the panel board on a higher floor.

The incident was investigated by the US Chemical Safety Board (CSB). In their report, CSB pointed out that the company's learning culture may not have been sufficiently strong. What other culture gaps might have contributed to this incident?

Did the PHA team fully understand and act on the hazards and risks of the emergency procedure to drain hot, reacting VCM from the bottom valve using an "emergency air line"? Did operators feel any sense of vulnerability that they might open the wrong valve when using the emergency air line? Why was the emergency air hose frequently used to drain the reactor when cleaning? Did this represent deviance that became normalized, or was there a gap in the imperative for safety that encouraged operators to defeat interlocks?

Understand and Act Upon Hazards/Risks, Combat the Normalization of Deviance.

E.28 How Many Explosions Does It Take to Create a Sense of Vulnerability?

> Actual Case History

In 2011, 3 explosions involving iron dust occurred over less than 4 months in a plant that manufactured finely divided iron powder (Ref E.2). In the previous 12 years, local firefighters responded to a total of 30 iron dust fire and explosion incidents at the site, although the 2011 explosions were more severe.

In the first explosion of 2011, iron dust was suspended in air by the jerky motion of a malfunctioning bucket conveyor. In the second, iron dust was suspended in air when a piece of equipment was struck with a mallet to drive a gas line into a fitting. In the third, a leak presumed to be from a nitrogen line turned out to be from a hydrogen line below the floor. In removing the access cover, a spark was created, causing a hydrogen explosion, which in turn created a pressure wave that suspended iron dust on nearby equipment. In all three cases, the iron powder flashed and exploded, the last one leveling the building. In all, 5 workers were killed, 3 were injured.

In all three 2011 cases, ignition took place nearly immediately, indicating the abundance of ignition sources. In 2 of the 3 cases, dust was suspended by mechanical action that could have been avoided, and in the third case, the mechanical action that ignited the hydrogen could have suspended dust if it had been present. In the third case, the hydrogen explosion lifted dust that had settled on many surfaces, causing the secondary explosions. It is only luck that prevented secondary explosions in the first two incidents.

The CSB in their report highlighted *poor understanding of the hazards and risks* of iron powder on the part of the company, and questioned the understanding of local, state, federal, and insurance inspectors. What other culture gaps might have contributed to this incident?

The explosiveness of iron dust is well-documented in the literature and was clearly stated in the plant's Safety Data Sheets. Yet, plant personnel tolerated a dusty workplace, did not take pains to prevent mechanical actions that could suspend dust, and tolerated 2-3 minor explosions per year. Did the tolerance of a dusty dirty environment represent the absence of *an imperative for safety?* Did a gap in performance occur because workers and management were not *communicating openly* about the hazards?

What factors led the facility to *normalize deviance* to the extent that they would think that 30 incidents in 12 years (or even one in one year) could be accepted as business as usual? Why did workers tolerate being in such a hazardous workplace? Did they feel *empowered* to improve the safety of their workplace? Employees reliably wore flame resistant clothing in the plant. However, the clothing did not provide significant protection to workers when the fires and explosions occurred. Was the use of fire resistant clothing part of a pattern of many factors used to dismiss workers' *sense of vulnerability*?

Establish an Imperative for Safety, Provide Strong Leadership, Maintain a Sense of Vulnerability, Understand and Act Upon Hazards/Risks, Empower Individuals to Successfully Fulfill their Safety Responsibilities, Combat the Normalization of Deviance.

E.29 Disempowered to Perform Safety Responsibilities by "Omniscient[1]" Software

Actual
Case
History

A plant sustained a small leak on the process side of a heat exchanger. Action was quickly taken to repair it, but during the shutdown, the coolant dropped the exchanger temperature dangerously low, embrittling the metal. As the process restarted, the heat exchanger ruptured, releasing a flammable vapor cloud. The vapor cloud traveled 170 meters before finding an ignition source. The massive gas cloud exploded and then caught fire, killing 2 workers and injuring 8. Because the plant was the sole supplier of natural gas to the region, the entire region had no gas for cooking, and factories employing 250,000 workers were left idle.

A corporate audit of the plant conducted just 6 months before the incident declared that the plant's process safety management system was in order. However, the incident investigation team found (Ref E.3) significant deficiencies in process hazard analyses, training, documentation, workforce involvement and communication, and management oversight.

The Royal Investigation Commission noted that the company had a world class computer-based system to manage its process safety programs, but concluded that the company's use of it was flawed in that personnel over-relied on checking the boxes specified by the system rather than assuring actual safety, effectively failing to *empower individuals to successfully fulfill their safety responsibilities.* What other culture gaps might have contributed to this incident?

What culture factors led the PHA team to fail to *understand the hazards and risks* they were evaluating and develop insufficient actions? Was failure to *ensure open and frank communications* and

---

[1] The word omniscient is used here in its literal sense, and does not in any way refer to the software company Omniscient Software Pvt. Ltd.

foster mutual trust the cause of the observed poor workforce involvement, communication, management oversight, and training?

Understand and Act Upon Hazards/Risks.

E.30 What We Have Here is a Failure to Communicate

Based on Real Situations

A plant producing an ingestible product from non-hazardous raw materials ruined a significant quantity of product by accidentally contaminating the product with facility wastewater. The error was detected while the product was still in the warehouse, so no customer was harmed, but if not caught many people could have been injured.

The process was implemented in equipment originally built for another process. The process tank had an overflow line that discharged to the facility sewer below the water level, which helped minimize odors related to the old process. Because the new process used vacuum charging of raw materials, the MOC process and the PHA identified that the overflow line needed to be removed. The minimal modification was performed via a simple maintenance work order.

After the process had been running for some time, the plant engineering department conducted an equipment audit and noticed that the overflow line, which was still shown on the P&ID, was missing. Over a weekend shutdown, they brought in contractors to replace the line. On Monday morning, the operator noticed immediately that the line had been replaced. He halted production until the line could be removed again by maintenance.

Several months later, Engineering reinstalled the line again, and the operator again noticed it and had it removed. Unfortunately, the third time the line was reinstalled, there was a new operator who did not recognize the change. When he pulled vacuum to charge the raw materials, he also siphoned wastewater into the tank. Due to the height of the overflow line, the amount of wastewater siphoned in was minor compared to the raw materials, so several batches were contaminated before the problem was detected.

The problem caused the loss of the business, and the Engineering manager retired suddenly. No formal incident investigation was conducted, but if it had been, what culture gaps might have been found?

This was not a process safety incident as we typically define it, but was the type of potentially high consequence low frequency type event that process safety prevents. What caused the significant gap in *open and frank communication* between operations, maintenance, and engineering, which undoubtedly extended to traditional process safety situations in the plant?

When engineering reinstalled the overflow line the second time, why did the operator or his management not discuss this with Engineering? When Engineering saw that the line had been removed again, why did they not discuss this with production? Did they not feel *empowered to successfully fulfill their safety responsibilities?*

Was *mutual trust* not fostered between Engineering and Production? Were these two groups only focused blindly on their respective duties, or was there a long-standing inter-departmental conflict?

Understand and Act Upon Hazards/Risks, Empower Individuals to Successfully Fulfill their Safety Responsibilities, Ensure Open and Frank Communications, Foster Mutual Trust.

E.31 Becoming the Best

Actual Case History

A manufacturing plant was struggling with their relationships with the environmental agency and their local community. Process deviations would frequently cause process emissions to spike, leading to significant odors offsite and exceeding the permitted level. The process would also periodically generate significant noise at a decibel level and frequency that was very irritating beyond the fence line.

One Earth Day, the plant dutifully held an open house to show off their state-of-the-art facility and show how they had all but eliminated their process and office waste. The neighbors were not as interested in this and their questions quickly turned to the site's odor and noise. One neighbor asked, "What is happening in the plant when emissions increase?" The technical manager hesitated, and then with encouragement of the plant manager explained that it happened when the gas rate was high. The neighbor then asked the same question about noise. The technical manager explained that it happened when the gas rate was low.

Another neighbor then asked, "So if you know why there are the odors and noise, why not avoid those conditions?" The technical manager explained that they were trying to, but it was not a direct correlation and they had not yet figured it out, but they would keep looking for a solution.

The meeting ended amicably, and a few days later, the technical manager received a call from the neighbor saying, "Your gas rate is low." The technical manager signaled through his office window and the operator turned up the gas. The neighbor, noticing the immediate reduction in noise replied, "Thank you." Over the following months, with neighbor input, the plant got the gas rate under control and achieved the lowest rate of emissions In the company.

This example shows how the linkage between culture and the PSMS element of stakeholder outreach. What positive culture attributes did the plant manager exhibit in being open with the community?

Establish an Imperative for Safety, Provide Strong Leadership, Ensure Open and Frank Communications, Foster Mutual Trust.

E.32 High Sense of Vulnerability to One Dangerous Material Overwhelms the Sense of Vulnerability to Others

> Actual
> Case
> History

A facility was restarting operations following a turnaround for replacement of a pressure vessel and a major control system upgrade. During start-up, a runaway chemical reaction occurred inside the pressure vessel, causing the vessel to explode violently. Untreated residue and highly flammable solvent sprayed from the vessel and immediately ignited, causing an intense fire that burned for more than 4 hours.

The fire was contained inside the unit by the plant fire brigade with assistance from local volunteer and municipal fire departments. Shrapnel from the explosion flew in the direction of a day tank containing a highly toxic chemical, but was stopped by protective shielding placed for this purpose.

Two employees who had been dispatched from the control room to investigate an unexpected pressure rise were near the residue treater when it ruptured. One died at the scene; the second 41 days later. Six volunteer firefighters and two contractors working at the facility were treated for possible toxic chemical exposure. More than nearby 40,000 residents, including students at the adjacent university, were ordered to shelter-in-place for more than three hours as a precaution.

The investigation team determined that the runaway chemical reaction and loss of containment of the flammable and toxic chemicals resulted from deviation from the written start-up procedures, including bypassing critical safety devices intended to prevent such a condition.

Other contributing factors included an inadequate pre-startup safety review; inadequate operator training on the newly installed control system; an unevaluated temporary change; and insufficient technical expertise available in the control room during the restart. Poor communications during the emergency between the company incident command and the local emergency response agency confused emergency response organizations and delayed public announcements on actions that should be taken to minimize exposure risk.

In managing the crisis, the company reported that "no toxic chemicals were released because they were consumed in the intense fires." While a reasonable assumption, investigators found that air monitors placed near the unit to detect toxic chemicals were not operational at the time of the incident, so this could not be confirmed. Management also attempted to prevent public access to information about the accident by asserting that the facility was covered by regulations related to sensitive security information. This assertion was determined by the governing authority to be without basis. Management later acknowledged that this was done due to limit the potential outcry related to existence of the highly toxic chemical at the plant.

The investigators provided numerous examples of the company using good engineering and operating practices to protect against releases of the highly toxic chemical, including reducing inventory, locating the main storage tank underground, shielding the above-ground day tank, and providing a dump tank if necessary to rapidly empty the day tank and associated piping. And in fact, these procedures were effective and well-managed. While investigators did not examine culture, readers can deduct from the investigation report that the process safety culture related to this unit was robust.

However, it is not clear that the PSMS and culture was functioning as well in the adjacent unit. If the investigators had examined culture, what potential culture gaps might the investigators have considered exploring?

Did an extra high *sense of vulnerability* from the highly toxic chemical reduce company employees' *sense of vulnerability* related to other chemical and processes?

Modeling showed that if a larger piece of shrapnel had struck the shielding around the toxic chemical day tank, the shielding might not have been able to protect the tank. Despite the good engineering work to protect the day tank, did the company fully *understand and act on the risk* related to the impact of incidents in nearby equipment?

Does bypassing interlocks, skipping the pre-startup safety review, and paying insufficient attention to operator training on new equipment suggest a *weak imperative for safety* in that unit?

What could the company's attempt to shield the incident from public scrutiny indicate about the *communication* and the level of *trust* that existed between the company and the public? Did that extend to workers?

Ensure Open and Frank Communications, Foster Mutual Trust, Combat the Normalization of Deviance.

E.33 Not Empowered to Fulfill Safety Responsibilities? Maybe You Were All Along

Actual Case History

An engineer visiting a plant for the first time arrived to find his access to the gate house blocked by an arriving tank truck. He watched as the truck crept onto the scale, and noticed with amusement that the truck's wheelbase was 15 centimeters too long to fit on the scale.

His amusement increased when the truck backed up and the guard placed a railroad tie, conveniently fitted with handles, on the front end of the scale. The driver than accelerated and quickly applied the brakes, stopping expertly with his front wheels hanging over the front of the railroad tie, suspended above the ground. With the back wheels now on the scale and the weight of the front wheels transferred to the railroad tie, the truck was weighed. The truck then rolled off the railroad tie, the railroad tie was removed, and truck continued to the unloading area.

As the truck pulled away, the engineer read the label on the tank car that he had not noticed earlier: "Chlorine, liquid". He then noticed the relief valve atop the end of the tanker, and realized that if the driver was any less expert, liquid chlorine could have sloshed with enough force to open the relief valve, not far from where he was standing. While he might have been able run away, the guards would have been trapped in their building.

The engineer proceeded to the Plant Manager's office and inquired about the situation. "We worry about this every day," the Plant Manager said. "We'd prefer to switch to rail transport, but our chlorine usage is not enough. And corporate will never approve a new scale. So, we are stuck with it. Luckily, it is the same driver every time and he is very good."

The engineer asked if the plant manager had ever requested the new scale, considering the potential consequences. He had not. "Why don't we try?" the engineer suggested. The plant manager wrote an appropriation request, noting the risk caused by the too-short scale and inviting the regional director to observe the weighing of the truck. The new scale was approved in the next budget cycle, and installed soon after.

This example shows how employees may have greater *empowerment* to address process safety issues than they believe they have. What factors could have contributed to the Plant Manager incorrectly believing he could not address this risk?

The engineer was certainly not the first company employee to visit the plant. What factors could have prevented other company visitors from mentioning it?

Establish an Imperative for Safety, Maintain a Sense of Vulnerability, Understand and Act Upon Hazards/Risks, Empower Individuals to Successfully Fulfill their Safety Responsibilities, Combat the Normalization of Deviance.

E.34 Normalization of Ignorance

A company was created by chemist and chemical engineer to manufacture a high value chemical. While both were experienced researchers, neither had experience developing, designing, and operating processes involving chemical reactions. They hired three recent chemical engineering graduates to operate the plant.

The plant operated without incident for three years, although there were several batches with significant exotherms that were difficult to control. One day, a more serious exotherm took place. Suspecting a problem with the cooling system, the owner/engineer and an operator/engineer went to investigate.

Before they could determine the problem, the reactor vessel burst, killing both and damaging property in a 400-meter radius. Debris from the blast was found more than 1.5 km away. The CSB investigation team (Ref E.4) found that no reactive chemical testing had been conducted during the design of the plant, the relief valve was not sized to handle the runaway reaction case, and the cooling system was significantly undersized and had no back-up.

CSB also noted that none of the company's employees had any knowledge of or exposure to reactor design or reactive chemical hazards. They noted that chemists and engineers are taught about preventing reactive chemical hazards primarily as in-company training in larger companies having a reactive chemical program; few degree programs addressed this subject. Noting an overall academic culture that neglected process safety, CSB recommended that the undergraduate chemical engineering curriculum requirements be changed.

While this was clearly a wise recommendation, what other culture factors might CSB have explored in this investigation?

What led the company to accept large exotherms that did not run-away, observed on previous batches, as a success instead of an opportunity to learn from a near-miss? What led the owners to think that an inexperienced chemical engineer would be a better choice for running the process than an operator experienced in running reactions on the industrial scale? Did the high price commanded by the product make the owners more willing to tolerate a sloppy process?

Establish an Imperative for Safety, Maintain a Sense of Vulnerability, Understand and Act Upon Hazards/Risks, Defer to Expertise.

E.35 Spark and Air Will Find Fuel

Actual Case History

A wastewater treatment plant provided a metal roof to shade a plastic methanol storage tank to reduce solar heating of the tank. After the roof was damaged in severe weather, facility management decided to remove it. Two workers in a man-lift basket cut the roof into sections, and each section was lowered to the ground by a crane. The work proceeded over several days.

While cutting a section near the tank vent, a spark ignited vapors coming from the vent. The fire flashed back through the vent's malfunctioning flame arrestor, igniting vapors in the headspace. The pressure from the resulting fire separated most of the connections from the tank, including the flame arrestor, the level transmitter, a level switch, and inlet and outlet piping. The flat bottom of the tank also bulged and pulled off its foundation. Flaming methanol vapor discharged from vapor-space connections, burning the two workers and the crane operator. One burned worker fell to his death, one died from his burns, and the third was hospitalized for more than four months.

The CSB investigators (Ref E.5) determined that the facility's engineering contractor had selected the wrong materials of construction for the tank and the flame arrestor. Consequently, the aluminum flame arrestor had corroded to the point it no longer functioned and the plastic tank could not withstand the pressure and stresses of the internal and external fire.

The investigators further discovered that the facility did not have a permit-to-work system, that it was seriously overdue on equipment inspection, and that its frequency of safety training had been steadily decreasing over the prior eight years. Based on interviews, the last training involving methanol hazards had occurred twelve years earlier.

In its investigation report, the CSB made recommendations to regulatory agencies, standards organizations, and the engineering company that installed the methanol system. However, it made no recommendations to the facility, which is ultimately responsible for worker and process safety. Which culture factors could CSB have explored in this investigation?

Did facility management and workers understand the hazards and risks of its processes? What caused the decrease in training frequency? Was the imperative for safety weakening?

High consequence scenarios related to the intended project are easy to imagine. Hot cuttings could partially or completely melt through the plastic roof of the tank or piping. Methanol venting from the tank as the sun heats it could ignite from cutting sparks. A cut-off roof section could be dropped edge-on and slice through the tank or piping. What caused workers and management to not think about any of this? Or if they thought about it, what caused them to not act to protect against these seemingly likely deviations?

Maintain a Sense of Vulnerability, Understand and Act Upon Hazards/Risks.

E.36 Operating Blind

A worker was lining up valves to transfer kerosene and gasoline from one terminal to a neighboring

| Actual |
| Case |
| History |

terminal operated by another company. During the process, he began to change the position of a figure-eight type line-blind valve. Unfortunately, block valves upstream of the blind had been opened out of sequence. As he swung the blind, a jet of gasoline sprayed out at high volume. The worker was unable to stop the release, and was soon overcome by fumes.

A supervisor attempted to rescue the worker, but he too was overcome with fumes and barely escaped. A second worker also attempted rescue and was also overcome. A third worker normally on site was offsite for personal reasons, leaving no one at the site to initiate further control actions. By the time response personnel arrived, vapors from the leak had engulfed the entire site. Recognizing the danger of explosion, they retreated.

One hour and fifteen minutes later, the vapor cloud exploded, creating a fireball that engulfed entire site and broke windows in the surrounding community up to 2 km away. The ensuing fire soon spread to all the other tanks on the site and continued to rage for eleven days. Due to the scale of the fire, responders decided to allow the fire to burn itself out rather than try to control it. Ultimately six workers and five in the community lost their lives.

The investigating commission (Ref E.6) determined that the accident was caused by valves being operated out of sequence, and was exacerbated by the absence of a remote isolation valve and/or remotely operated shut-off. The commission noted that there were no operating instructions for making the transfer, leaving the procedure up to the operators who were not well trained in this procedure.

While the commission did not comment specifically on company process safety culture, several recommendations show they were clearly thinking about it. Among the recommendations, the commission recommended creating an independent process safety function reporting to the CEO and that line management practice conduct of operations to ensure that all process safety functions are carried out. What other culture factors could the commission have considered?

Did the fact that the operation involved a transfer from one company to another create a "not my problem" attitude? The commission noted a lack of training in the procedure. What was the general status of training in the facility? Were workers trained to recognize and control hazards and risks? Did they take part in "man-down" drills? What was the current focus of corporate process safety efforts? Were employees *empowered to fulfill their safety responsibilities*?

Provide Strong Leadership, Maintain a Sense of Vulnerability, Understand and Act Upon Hazards/Risks, Empower Individuals to Successfully Fulfill their Safety Responsibilities, Combat the Normalization of Deviance.

E.37 Playing Jenga® with Process Safety Culture

Based on Actual Situations

Jenga® is a Parker Brothers strategy and skill game. Players construct a tower of blocks, and then take turns removing a block from the middle of the tower and adding it to the top. The last to successfully remove a block without toppling the tower is the winner.

A Vice President of Operations of a company, a long-time employee well-steeped in the company safety culture, noticed that process safety leading indicators and near-miss metrics were beginning to trend negatively across the company. While the trend was not strong, the Vice President called a global meeting of safety and operations leaders that all were required to attend. The purpose of the meeting was to develop an action plan to ensure the unfavorable trend did not continue and the company could get back to its previous performance.

Not long afterward, the company began shifting the focus of its business. Coincidentally, the Vice President of Operations retired. The new Vice President was challenged by the business shift and may have been distracted from the predecessor's process safety action plan. After a few more years and more staff changes, the company experienced a cluster of major and minor incidents involving injuries and fatalities that would have been unheard of just a few years later. Over the next few years, incidents began cropping up at many sites, and to many, the company's process safety culture appeared to have collapsed.

Could the company's culture have been toppled like a Jenga tower by the removal of one strategic piece? Or, continuing the analogy, can culture survive many small weaknesses until critical failure? In other words, was the departure of the Vice President the critical factor, or had the culture become so weakened that it would have collapsed even with that Vice President's leadership?

What was the true state of culture when the metrics trend first started to go negative? Could the culture have already collapsed, and could the absence of *frank and open communication* have prevented the well-meaning Vice President from knowing about it until it was too late?

Did the company truly have an *imperative for safety*, or was its reputation built on a few very visible safety champions? Could the *imperative for safety* been focused on occupational safety and not enough on process safety? Could the Vice President's successors have talked-the-talk about process safety, but not exhibited "Felt Leadership?"

Establish an Imperative for Safety, Provide Strong Leadership, Maintain a Sense of Vulnerability, Understand and Act Upon Hazards/Risks, Ensure Open and Frank Communications.

E.38 Failure of Imagination?

In February 1967, an electrical fire within the crew capsule of the Apollo 1 spacecraft killed all three astronauts as they conducted a simulated launch

> Actual
> Case
> History

drill on the launch pad. The investigation determined that the oxygen atmosphere in the capsule caused a minor electrical short to accelerate into a significant fire. The crew and launch attendants outside the capsule tried to open the hatch, but the combustion gasses had raised the cabin pressure enough so that the inward-swinging hatch would not budge.

Before the incident, Apollo astronauts had expressed many concerns about their new spacecraft, including a significant amount flammable nylon webbing throughout the crew cabin. The investigation board noted that NASA had failed to identify flammability hazards so that they could have been addressed.

During the investigation hearings, an astronaut termed the failure to connect flammables plus oxygen to fire was a "Failure of Imagination." Of course, it was not a failure of imagination because the Apollo 1 crew had imagined it – and have even complained about it.

If the crew complained about a safety problem was there was an *understanding of hazards and risk*, but a failure at some level of the organization to *act on these hazards and risks*? Were the crew aware of the hazards but other astronauts failed to imagine it? If so, was there a gap in *open and frank communication*? Did the others not have the *same sense of vulnerability*, or did they not trust their colleague's judgment?

Establish an Imperative for Safety, Maintain a Sense of Vulnerability, Understand and Act Upon Hazards/Risks.

E.39 Playing the Odds

Actual Case History

A young engineer overseeing his first plant trial batch was discussing the first step of the operating instructions with a 35-year experienced operator.
"We can skip the inerting step," the operator said. "That will save us some time to have coffee and eat those nice donuts you brought for me and my buddies."

The engineer shook his head and explained patiently that it was necessary to inert the reactor, because otherwise the flammable atmosphere could ignite, especially because the solvent was not being fed through a dip-pipe. "Yeah, I've heard of that," the operator said, "but take it from me, it is a waste of time to inert the reactor because 9 times out of 10 it does not explode."

"Uh, let's have that coffee and talk about it," the engineer said. They went into the breakroom, took their coffee, and sat across the table from each other with the box of donuts between them. The engineer reached for a donut. "The thing is," he said, "if it doesn't explode 9 times out of 10, then it does explode that other one time. I don't know about you, but my goal is this." He held the donut up in front of him, showing the operator the big sweet 0.

The operator grabbed the donut and stuffed it in his mouth. After washing down that donut with a gulp of coffee, he put 2 more donuts in his pocket, left the breakroom and started inerting the reactor.

The operator appeared to understand the hazard and possibly even the risk. If so, did he need to have it explained to him again? Or did he need something else? Did the operator frequently skip other safety steps in procedures? Was this normal behavior within the plant? Should the engineer have questioned the Plant's *imperative for safety*?

How did the engineer convince the operator? Was it through a logical argument? Establishing *mutual trust*? Or was the operator testing the engineer's *leadership*?

Establish an Imperative for Safety, Provide Strong Leadership, Maintain a Sense of Vulnerability, Understand and Act Upon Hazards/Risks, Defer to Expertise, Combat the Normalization of Deviance.

E.40 Shutdown and Unsafe

A distillation column in a unit producing an aromatic nitro compound exploded, shooting the top half of the column and other debris up to half a kilometer away. Three employees were cut by glass from a blown-in window, and some large fragments came close to storage tanks of flammable and toxic materials but fortunately caused no damage.

The investigation team (Ref E.7) found that the plant had been in an extended shut down, which included shut down of the plant boiler. However, when the boiler was restarted, steam began flowing to the column reboiler at a slow rate through a leaking manual bypass valve. The liquid left in the bottom of the column began to heat slowly and eventually began to generate vapor. The vapor rose through the column and condensed on an upper tray, causing a level alarm on the tray to sound. Operators, not expecting the column to be running, silenced the alarm and ignored it.

The reboiler continued to heat, and eventually a runaway reaction began in the reboiler. The boiling rate increased significantly, overcoming the capacity of the narrower upper part of the column. The relief valve, designed for fire case, not for runaway reaction, was unable to relieve the pressure. A small breach formed first, then the top of the column burst from the lower section and flew off.

The investigators found numerous errors in the PHA, and operating procedures. If investigators had investigated the plant's process safety culture, what might they have found?

Operators dismissed the tray high-level alarm instead of questioning why it sounded when the column should have been shut down. Did they take the shutdown condition as "permission" to dismiss their *sense of vulnerability*? Did they not understand that the nitro compounds left in column during the shutdown still represented a risk that needed to be managed?

Bypass valves tend to be used only occasionally. Was maintenance of this valve dismissed because it was considered insignificant? Did the PHA team *understand and act on the potential hazard* caused by the failure of this valve? The reboiler temperature began to rise immediately after the boiler was brought on line, but was not noticed because it was not being checked by operators during the shutdown. What was management's role in not continuing to monitor key process variables for equipment containing a hazardous material?

Maintain a Sense of Vulnerability, Understand and Act Upon Hazards/Risks, Combat the Normalization of Deviance.

E.41 Who, me? Yeah, you. Couldn't be. Then who?

Actual
Case
History

A fire and deflagration explosion happened at a liquid waste injection well site when basic sediment and water (BS&W) from two natural gas wells arrived contaminated with hydrocarbons. While the BS&W was being unloaded, the hydrocarbons ignited, causing a deflagration and subsequent pool fire. Two workers were killed, and three others were injured.

The liquid waste injection site was owned by one company, the natural gas well by a second company, and the waste was being transported by a third company that was responsible to extract the BS&W from the storage tank using a vacuum truck. The gas well owner was to notify the waste hauler which tank to extract from and the volume to extract, expressed either in inches of outage or barrels as appropriate. The waste hauler was then to extract that quantity from the bottom of the tank, checking to ensure that little or no hydrocarbon, which floated on the BS&W, was removed.

Each driver had a different method for measuring the amount removed during the vacuum operation and or detecting the BS&W to hydrocarbon interface. The official amount transported was determined only by the owner by level difference after the hauler departed the well site. The hauling company and owner both clearly understood that no hydrocarbon should be removed from the tank during the extraction operation, but no check of the extracted material was made to confirm this before transporting.

On the day of the incident, investigators concluded that a significant amount of hydrocarbon was unintentionally extracted. When the truck was being unloaded at the liquid waste injection site, hydrocarbon vapors from the tank were ignited, most likely from the idling truck engine. In the ensuing fire, the truck valve opened, draining additional BS&W and hydrocarbon to the unloading pad. This hydrocarbon formed a pool fire that took nearly an hour to extinguish.

The investigators (ref E.8) noted several management system failures as well as regulatory gaps that contributed to the incident. The investigator further noted that the industry generally recognized BS&W as non-hazardous, and that while some in the industry recognized that hydrocarbon that could be present in extracted BS&W could be flammable, the majority did not. This difference could simply one of terminology: "flammability" is defined as having a flashpoint below 100 °F while liquids with flashpoints not too far above that temperature might can burn and can still ignite readily, especially if warmed.

Relying on regulatory definitions when they are not accurate, and denial of hazards are clear signs of a weak *sense of vulnerability* and a *weak imperative for safety*. What other culture issues might have existed in this situation?

The well owner clearly *empowered* the hauler to verify the absence of hydrocarbon in the extracted BS&W, and this would seem to be a culture positive. Likewise, the waste injector *empowered* the hauler and the well owner to make this verification. However, where does empowerment end and become abdication of responsibility, a culture negative? Did that happen in this case?

The crude and inconsistent methods of determining the BS&W to hydrocarbon interface was prevalent throughout the industry. This suggests that hydrocarbon could be present in many such BS&W pump-outs. If some in the industry considered this to be a hazard and others did not, were there barriers to *openly and frankly communicating* this concern? Were there barriers to *learning and advancing the culture*? How did the transfer of safety responsibility between the three companies involved in the operation create other opportunities to weaken safety culture?

Maintain a Sense of Vulnerability, Understand and Act Upon Hazards/Risks, Ensure Open and Frank Communications.

E.42 Blindness to Chemical Reactive Hazards Outside the Chemical Industry

Actual Case History

A plastics extrusion plant suffered a multiple fatality incident when workers were attempting to open a waste plastic tank to clean it. The vessel pressure gauge showed no pressure in the vessel, but the gauge had become blocked with plastic and did not show the actual pressure in the vessel. After half the bolts fastening the vessel cover had been removed, the cover flew off, killing the three workers. The cover also severed hot oil lines, leading to a fire that took several hours to extinguish.

Investigators (Ref E.9) discovered that the plastic had a reactive chemical hazard, an exothermic decomposition reaction at hot temperatures. As the plastic in the catch tank cooled on the outside, the plastic in the center remained hot and molten, allowing the decomposition reaction to continue to build pressure, while the solid plastic outer shell shielded the pressure gauge from detecting the high pressure in the tank.

Investigators found that while the company was not aware of the plastic's decomposition reaction, the company had had more than 20 minor incidents or near misses over nearly 10 years that provided many hints of the existence of this hazard. While some of the minor incidents or near misses could be explained individually by labeling them as "process fires", some of the fires occurred in environments without oxidant or ignition source. The plant launched a process fire prevention program, but it was unsuccessful and abandoned.

What culture factors were involved in this incident?

It is not unusual for facilities that handle materials that are not considered "chemicals," such as petroleum, plastic, food, etc., to neglect the potential for chemical reactivity hazards. Which culture elements need to be strengthened in such companies to help them evaluate potential hazards that might be thought to be "outside the box" for them, but really are not?

Are there other examples for such "industry blindness" to hazards that are considered "the other industry's problem?"

This appears to be an extreme example of the *normalization of deviance*. Why ultimately did the reactivity hazard issue get normalized?

The process fire program was abandoned without considering some other way to address the fire related incidents. Did workers accept fires as normal business, or did they complain to deaf ears? Did management try to address fires but could not cooperation from workers? Did anyone check external sources for help with the issue?

Maintain a Sense of Vulnerability, Understand and Act Upon Hazards/Risks, Combat the Normalization of Deviance.

E.43 Dominos, Downed-Man "Nos"

Unfortunately, this case history involves a lot more than one fatal incident, and may number in the hundreds, thousands, or more. It proceeds like

> Based on
> Real
> Situations

this. First, there is a release of some toxic or asphyxiant, either noticed or unnoticed. A worker becomes exposed, and collapses. Another worker, all too often a close friend or relative, sees the person down and rushes to help, only to be overcome and collapse also. Then, sometimes, comes a third, a fourth...

Clearly it is possible that a worker could collapse from heart attack, stroke, illness, or dehydration. But we who work in hazardous material industries know deep down that if we are in a hazardous material facility, toxic or asphyxiant exposure was more likely to cause a person down. We know we should pull the alarm and notify the emergency responders. We know that the situation must be assessed wearing respiratory protection. But the person down is our friend or our family member, and they need help. How can we make sure we do the right thing? How can culture help us overcome this problem?

Is there a gap of *trust* that the proper emergency response can be done on time? How can we ensure this trust? Is there a gap in *vulnerability*? Do we dismiss the possibility that toxics or asphyxiants could be released, and therefore ignore that possibility when we react? Do such incidents occur without warning? Do we see warning signs before they ever occur, but do not feel a sufficient *imperative for safety* to prevent them in the first place?

This type of incident happens often enough to suggest that we are not learning from it and advancing our ability to prevent it. How can we learn and implement learnings more effectively?

Establish an Imperative for Safety, Maintain a Sense of Vulnerability, Understand and Act Upon Hazards/Risks, Learn to Assess and Advance the Culture.

E.44 Mr. Potato Head Has Landed

Actual
Case
History

A group of balloonists were participating in a hot-air balloon festival featuring balloons designed to look like cartoon characters and toys. The wind carried them over a large chemical complex, where they found themselves caught in the rising column of warm air coming off the plant's cooling towers. While the updraft did not affect the balloons' ability to float, it did prevent them from drifting with the wind. The plant could not shut down its cooling towers, and the balloonists only had a finite supply of fuel, so there was no alternative: the hot-air balloons had to be brought down inside the plant. (Ref E.10)

The balloonists and plant personnel had to guide the balloons down in an orderly fashion between buildings, pipe racks, stacks, flares, and ponds without damaging the balloons or plant equipment, and taking care that the balloons' burners did not trigger any fires. Teams of employees spontaneously coordinated with the plant emergency management team to choose landing sites, guide balloons to them, deflate the balloons, and move them to clear the landing sites for the next balloons' descents.

All balloons were landed safely, with no damage to the plant and no injuries, even to Mr. Potato Head. What positive safety culture dimensions did the plant demonstrate in this unusual situation?

It is easy to imagine plant workers laughing at the balloonists being stuck over the plant, or even for them to view their predicament as a special show just for them. What culture attributes led workers to quickly understand this was a potentially dangerous situation?

Even though plant workers took the situation seriously, they laughed about the situation as they went about their rescue work. To what degree is good-natured humor an indicator of a strong safety culture?

This was clearly not the kind of emergency response that anyone in the industry would plan for or train. Yet, the response was executed flawlessly. What culture dimensions help made this possible?

Establish an Imperative for Safety, Understand and Act Upon Hazards/Risks.

E.45 Sabotage, Perhaps. But of the Plant or the Culture?

Actual Case History

A massive gas explosion at a government-owned refinery killed 7 workers and 40 people offsite including 35 members of the National Guard and their family members (Ref E.11). Security footage and eyewitness accounts suggest that a gas leak began in the morning, but was hard to detect due to heavy rain and mist, except for occasional whiffs of rotten eggs. Just before midnight, the rain eased, and the gas leak became more apparent. Investigators believed that the explosion was initiated by an offsite truck starting its engine.

Investigators discovered that the bolts on a pump head had worked loose, enabling the gas to leak out. This type of failure can happen on excessively vibrating equipment, and with the plant's reputation for a weak mechanical integrity program, this could be taken as the incident's root cause. Inspection of the bolting showed that seven of bolt studs were only partially threaded into the pump body. Some were over-stressed, and some of the nuts and bolt-ends had bite marks left from improper use of a pipe wrench.

However, the investigators concluded that these bolting issues were evidence of sabotage conducted, they hypothesized, by a government opposition group. Considering the weak security of the plant and the proximity of the compressor to the property boundary, saboteurs could have entered the plant and loosened the bolts. Regardless of the true cause, culture was clearly at play in this incident.

No person or group took responsibility for sabotage, which normally occurs. Could the sabotage theory have been advanced to enable workers, managers, and the government as an excuse for not *fulfilling their safety responsibilities*?

If the cause was not sabotage, then the pump head had clearly been short-bolted during a prior maintenance activity, perhaps accepting the short-cut rather than cleaning out and re-tapping the bolt holes in the valve body. Were there other examples of *normalization of deviance* in plant maintenance activities?

The plant circulated a survey asking employees whether they felt the incident was caused by sabotage or safety failure. Did employees feel compelled to select sabotage?

The plant was bordered closely on all sides by residences and businesses. How did the plant interact with the community on safety issues?

Maintain a Sense of Vulnerability, Understand and Act Upon Hazards/Risks, Combat the Normalization of Deviance.

E.46 This is the Last Place I Thought We'd Have an Incident

Actual Case History

An inorganic powder used as an oxidation catalyst was being isolated for disposal. The powder had been filtered from the reaction mixture and washed with clean solvent, and the solvent was being removed by sweeping the filter with warm inert gas through a chilled water condenser. During the drying cycle, an exothermic reaction occurred in the filter that damaged it. The mix of inorganic powder and organic solvent exited the filter and found an ignition source. The resulting overpressure caused some damage, and the fire was quickly extinguished by the fire suppression system. Fortunately, no injuries resulted.

The investigation team (Ref E.12) found that some years earlier, reactivity testing had identified a reaction between the powder and solvent that had a 24 hour "time to maximum rate" a few degrees below the drying temperature.

The drying had not always been run at a warm temperature. However, when drying was done at a cooler temperature, an unacceptable amount of solvent remained in the filtered powder. To obtain a drier cake, improving the occupational safety during pack-out, the plant increased the drying temperature. The MOC review concluded that the change was necessary to improve safety in packing out the cake. However, in conducting the MOC review, the original reactivity data were not fully considered. Instead, a new set of thermal tests on solvent-free powder were conducted, which was found to be quite stable.

When the investigation team reported its findings, a manager commented, "This is the last place I thought we'd have an accident." He explained that the site ran many highly energetic reactions, handled highly toxic chemicals, and distilled many volatile and flammable solvents. Surely if there was an incident on the site, he said, it would not happen in a filter.

What culture questions might the investigation team have considered?

Did an unbalanced *imperative for safety* lead to disregarding the original reactivity study (process safety hazard) so that the plant could address the solvent exposure issues (potential occupational safety hazard) during pack-out?

The technical team responsible for the process was unaware of the potential for reaction between powder and solvent. What barriers to *open communication* could have existed between the technical team and the owner of the process safety information? What other communication roadblocks might there have been?

Did the plant feel less *sense of vulnerability* for that operation than they should have because it was "merely" a filter, and if so, could that have contributed to the incident? How should a site ensure a sense of vulnerability for all processes, including ancillary/non-mainstream process hazards?

Establish an Imperative for Safety, Maintain a Sense of Vulnerability, Understand and Act Upon Hazards/Risks

E.47 References

E.1   Chemical Safety and Hazard Investigation Board, Investigation Report – Vinyl Chloride Monomer Explosion, Formosa Plastics Corp., 2007

E.2   Chemical Safety and Hazard Investigation Board, Investigation Report – Hoeganaes Corporation Fatal Flash Fires, 2011.

E.3   Dawson, Sir Daryl M., The Esso Longford Gas Plant Accident – Report of the Longford Royal Commission, June 1999.

E.4   Chemical Safety and Hazard Investigation Board, Investigation Report – T2 Laboratories Inc. Reactive Chemical Explosion, 2009

E.5   Chemical Safety and Hazard Investigation Board, Investigation Report – Bethune Point Wastewater Plant Explosion, 2006

E.6   Oil Industries Safety Directorate, IOC Fire Accident Investigation Report, 2009

E.7   Chemical Safety and Hazard Investigation Board, Investigation Report – First Chemical Corp. Reactive Chemical Explosion, 2003

E.8   Chemical Safety and Hazard Investigation Board, Investigation Report – Vapor Cloud Deflagration and Fire, BLSR Operating, Ltd., 2003.

E.9   Chemical Safety and Hazard Investigation Board, Investigation Report – BP Amoco Thermal Decomposition Incident, 2002

E.10 Lodal, P., Conference Introductory Remarks, CCPS International Conference, Scottsdale, AZ, USA, 2003

E.11 Compiled from news sources

E.12 Anonymous Personal Communication

APPENDIX F: PROCESS SAFETY CULTURE ASSESSMENT PROTOCOL

F.1 Introduction

The following questions that can be used to assess the status of the process safety culture in an organization. Some questions are intended to highlight evidence of a positive, while others help diagnose negative process safety culture.

Like any checklist or protocol, it is inherently incomplete. Answers to questions may prompt deeper investigation, and facilities may have cultural aspects that this protocol does not address.

Any symptom identified through this protocol whose impacts are severe or have resisted correction should be subjected to a separate formal analysis.

The questions in this protocol were derived from other sections of this book, particularly Chapter 4, which describes the relationship of process safety culture to each PSMS element, as well as from the references cited at the end of this section.

F.2 Culture Assessment Protocol

Establish an Imperative for Safety

1. Has the organization adopted a minimalist approach to PSMS applicability? A minimalist approach refers to a conscious effort to limit the PSMS boundaries only to the strict limits defined by any applicable process safety regulations affecting the facility and nothing else. This is sometimes referred to as a compliance-only approach. The use of a minimalist approach is usually an overt decision but may also occur because of all of the actual process safety risks have not been fully evaluated. For example, the inclusion of utility, support, and other systems that do not contain any process safety related chemicals but are critical to process safety. The failure of some

Essential Practices for Creating, Strengthening, and Sustaining Process Safety Culture, First Edition. CCPS. © 2018 AIChE. Published 2018 by John Wiley & Sons, Inc.

of these systems and equipment can lead to or contribute to a process safety incident, and hence deserve the same consideration as the main process systems and equipment that contain the hazardous materials of concern.

2. Is the PSMS driven by the "it cannot happen here? That is, serious process safety incidents are not possible or so rare that the PSMS can be designed and implemented with this philosophy as a basis.

3. Is the PSMS (particularly the performance of PHAs) governed by the "double jeopardy doesn't count" philosophy? Double jeopardy in this context means that more than one concurrent failure should not be considered a credible cause of a process safety scenario. Note that multiple latent (or unrevealed) failures in place waiting for a single triggering initiating event should not be treated as a double jeopardy situation.

4. Is the presence of other strong EHS related programs, such as environmental programs or achievement of OSHA's Voluntary Protection Program (VPP) Star status used to limit the scope or applicability of the PSMS? Note that VPP program inspections do not focus only on process safety but examine the full spectrum of health and safety programs in a facility.

5. Is the PSMS a detailed set of management system procedures that represents a "paper only" program that sits on the shelf, or has it actually been implemented and is it being used?

6. Are the scope and boundaries of PSMS extended to cover other hazards that are not covered explicitly by regulation but have been shown by incident history to represent significant process safety risks? This is an extension of the philosophy of not adopting a minimalist approach to process safety. An example of other process safety hazards are combustible dust hazards.

7. Is there a system in place that ensures an independent review of major process safety-related decisions? Are reporting relationships such that impartial opinions can be rendered?

8. Does the organization believe that MOC is important, and that changes cannot occur, however convenient they may be, or however simple and obvious they may seem without the appropriate review and authorization using the MOC process?
9. Is there a "shoot the messenger" mentality with respect to dissenting views, or raising process safety problems?
10. Are the decision makers technically qualified to make judgments on complex process system designs and operations? Are they able to credibly defend their judgments in the face of knowledgeable questioning? Do process safety personnel find it intimidating to contradict the manager's/leader's strategy?
11. Do production and protection compete on an equal footing when differences of opinion occur as to the process safety/safety of operations?
12. Has the staffing of key process safety positions been shifted, over the years, from senior levels to positions further down the organization? Are there key positions currently vacant?
13. Does management encourage the development of safety and risk assessments? Are recommendations for safety improvements welcomed? Are costly recommendations, or those impacting schedule, seen as "career threatening" if the person making the recommendations chooses to advocate them?
14. Is auditing regarded as a negative or punitive activity? Are audits conducted by technically competent people? How frequently do audits return only a few minor findings? Is it generally anticipated that there will be "pushback" during the audit closeout meetings?
15. Is safety and process safety a core value? Are the core process safety values are written down and stressed in training and other forums? Is there a company or facility document that describes process safety as a core value? Is there evidence, e.g., minutes of meetings and agendas for safety meetings or

other training and information that show that process safety is a core value?

16. Are process safety performance goals, objectives, and expectations included in performance contracts, employee goals and objectives, and discretionary compensation arrangements for line managers, supervisors, and workers?

17. Are the metrics or other means by which process safety performance is measured defined?

18. Do personnel report a pressure to maintain performance standards, potentially at the cost of safety?

19. Are there commitments to achieving performance goals that are greater than demonstrated for process safety goals?

20. Do operational pressures lead to cutting corners where process safety is concerned?

21. Is process safety improvement a long-term commitment that is not compromised by short-term financial goals?

22. Is there sufficient staff in relevant work groups (e.g., operations, inspection, or maintenance) to perform jobs safely?

23. Is the organization is preoccupied with safety and process safety, such that they can anticipate areas of potential failure and can cope and bounce back from errors when they occur? Do they exhibit a resilient nature? Resilience is defined as the ability of systems to survive and return to normal operation despite challenges.

24. Is process safety management an independent function in the organization? Does the main person responsible for process safety report to those who might have a conflict of interest with respect to decisions about the process safety impact on operations? Note: In smaller organizations this independence may be more difficult to achieve.

25. Are process safety resources are among the first budget line cuts during times of financial difficulty?

26. Is the process safety staff placed in the untenable position of having to prove that an operation is unsafe? Are those desiring

or advocating certain operations or conditions required to prove that those operations or conditions are safe?

27. Are the collection and analysis of process safety metrics treated as adversarial or punitive activities?

28. Are managers less strict about adherence to procedures when work falls behind schedule?

29. Does the tension between production and safety result in a slow and gradual degradation in safety margins?

30. Are shortcuts encouraged and rewarded to meet production or other goals?

31. Are rewards and incentives heavily weighted towards production outcomes?

32. Has the organization included inherently safer technologies considerations in its process safety program?

33. Have critical safe work practices (SWP) been designated as "Life Saving Rules" or "Cardinal Rules" (or similar designation for inviolable rules) for their application, with no tolerance for not doing them right every time? Implementing and maintaining a "Life Saving Rules" program requires that management enforce the rules consistently.

34. Does the imperative for process safety include understanding and accommodating, and sometimes influencing/advancing the related cultures of outside organizations that interact with a facility and affect its PSMS? These outsiders include contractors, regulators, unions, corporate staff, boards of directors, interest groups, community groups, and others.

Provide Strong Leadership

35. Does management have a firm understanding of risk and process safety in general, and accepts the identification of high-risk levels? Does organization senior management understand the technical aspects of process safety and how process safety requirements are interpreted for the site/company?

36. Is there stability of personnel in hourly or management positions? Has the turnover rate of facility managers, EHS managers, and process safety coordinators/managers been too rapid such that they do not have adequate time to learn their responsibilities? In particular, plant/facility managers who have been assigned for a relatively short period, primarily to fulfill a specific step in his/her career path, may not have the time, nor perhaps the inclination, to develop the knowledge required to adequately understand a performance-based program such as process safety nor to place a high level of priority onto the PSMS.

37. Is management visible, active, and consistent in its support for the PSMS and process safety objectives? Does this philosophy extend down through the ranks of middle management within the organization?

38. Do decisions about corporate-level initiatives, operations, financial performance, resource allocation, capital projects, personnel changes, compensation, and other aspects of operations visibly and tangibly demonstrate a commitment to process safety excellence?

39. Is there any confusion over who is responsible for what in the PSMS? Do job descriptions, performance goals documents, or the PSMS applicability or high-level policy/procedure describe the responsibility for each PSMS element, and its sub-parts?

40. Are PSMS accountabilities defined for each level of management and supervision in terms that are understood, and then enforced?

41. Is a significant portion of any variable pay plan contingent on satisfactorily meeting PSMS performance objectives? Note that there are both positive and negative aspects to safety-based compensation. Money incentivizes people very strongly. However, it can also result in adverse risk taking, not reporting incident or near misses to avoid losing a bonus, or other negative behaviors.

42. Are process safety performance and leadership significant considerations in career advancement and succession planning?

43. Has a company-level PSMS leader been designated? Is this designation made in writing and by title? Is this person technically competent in PSM? Does this person have sufficient positional authority to contribute meaningfully to the most significant process safety related decisions, and has that authority been influential?

44. Has a facility-level PSMS leader been designated? Is this designation made in writing (e.g., job descriptions, organizational charts, etc.) and by title? Is this person technically competent in process safety? Does this person have sufficient positional authority to contribute meaningfully to the most significant PSMS-related decisions, and has that authority been influential? Note: This position can be either full time or part time, depending on the size of the company, the number of facilities included in the PSMS, and the applicability and complexity of the company PSMS.

45. Are process safety issues identified dealt with by management and not just "filed".

46. Have process safety culture surveys and/or assessments been conducted, and have actions or priorities resulting from the survey been resolved? Have the surveys/assessments resulted in changes to the process safety culture?

47. Are the same process safety issues raised at each management meeting, but not resolved? Note: Management meetings can consist of a variety of forums, from daily production meetings, maintenance meetings, project meetings, process safety metrics, or other related meetings?

48. Does management resist taking responsibility for process safety concerns when they are faced with them?

49. Does management use empty slogans regarding process safety repeatedly?

50. Are process safety positions accorded equal status, authority, and salary to other operational assignments?
51. Is process safety an agenda item or is it considered at high-level meetings on a regular basis (not just after an incident)?
52. Is the process safety culture managed as a "program" or a separate and distinct activity for which a single person can be assigned responsibility or accountability, as one can be assigned to manage the PHA program at a facility? Note; While the assignment of "Coordinator" or even "Manager" to handle a program of special emphasis is sometimes a typical practice, the existence of a "PSM Culture Manager" or "PSM Culture Coordinator" is probably a sign that the organization and its management do not fully understand what "culture" means in this context.
53. Do managers and others cite the presence of other strong EHS related programs, such as environmental programs or achievement of OSHA's Voluntary Protection Program (VPP) Star status as evidence that the PSMS is also strong?
54. Are there competing values that dilute PSMS goals and objectives? Is there an EHS "flavor of the month" syndrome where the goals and objectives change depending largely on incidents and other events that have occurred?
55. Has a succession plan for the PSMS been developed? Is it up-to-date? Does the plan include the transfer of knowledge as well as ownership and responsibility?
56. Does management make periodic tours of the facility?

Foster Mutual Trust

57. Does the facility distinguish clearly between acceptable and unacceptable employee acts so that the vast majority of unsafe acts or conditions can be reported without fear of punishment?
58. Does the sharing of information that will reduce safety risks occur without fear of punishment?

59. Is there is a climate in which workers are encouraged to ask challenging questions without fear of reprisal, and workers are educated, encouraged, and expected to critically examine all process safety tasks and methods prior to performing taking them?

60. Are operational staff concerns not reported to management for reasons such as: staff are concerned that the report would get someone else in trouble; staff perceive that nothing would get done; employees feel that they may be deemed responsible for causing the issue?

61. Is blame apportioned or insinuated prior to or as a result of any incident investigation?

62. Can personnel report hazardous conditions without fear of negative consequences?

63. After a process-related incident, accident, or near miss, is management more concerned with assigning blame or issuing discipline than correcting the hazard?

64. Do personnel have confidence that a just system exists where honest errors can be reported without fear of reprisals? Do employees trust that the information they submit will be acted upon to support increased awareness, understanding, and management of threats to safety?

65. Are members of the organization afraid to challenge bad ideas when they are proposed?

66. Has a blame culture developed with respect to the process safety program? Blame cultures are characterized by: staff tries to conceal errors; personnel feel fearful and may report high stress levels; personnel are not recognized or rewarded and thus lack motivation; errors are ignored or hidden; management decisions tend to be taken without employee consultation; there is often a high staff turnover.

67. Has mistrust between groups or individuals caused severe differences in opinion and perception about the functionality of the process safety program such that little or nothing can be accomplished?

68. Are the results of audits and process safety metrics used directly in personnel performance reviews?
69. Are PSMS audits and process safety metrics programs treated or viewed as punitive activities?

Ensure Open and Frank Communications

70. Do both vertical and horizontal communications channels exist that encourage honest and open communications? This is generally accomplished both formally and informally. Formal communications methods consist of required reports and other written transfer of information, e.g., operator's logs, monthly process safety metrics reports, etc. Informal methods of communication can exist in many forms but are more unscheduled and ad hoc (i.e., as required) and perhaps the contents are not recorded in writing, e.g., a special briefing held before an operational task or maintenance job.
71. Are communications channels open? Are those reporting bad news or problems not at risk of being labeled as "non-team players" or of being ostracized? Is peer pressure is used to suppress these types of communications rather than to foster them? Is there a feedback loop where there is a response to subordinate's formal and informal communications?
72. Are lateral communications (e.g., between work groups or shifts) that are formal (e.g., shift turnover) and informal (e.g., radio communications between outside operators and the control room) effective?
73. Does management encourage communications that contradict pre-determined thoughts or direction? Are contradictory communications discouraged? Is the bearer of "bad news" viewed as a hero, or "not a team player?"
74. Does the organizational culture require "chain of command" communications? Or is there a formalized process for communicating serious concerns directly to higher

management? Is critical, safety-related news that circumvents official channels welcomed?

75. Are communications altered, with the message softened, as they move up the management chain? Is there a "bad news filter" along the communications chain?

76. Do management messages on the importance of safety get altered as they move down the management chain? Do management ideals get reinterpreted in the context of day-to-day production and schedule realities?

77. Are those bearing negative safety-related news required to "prove it is unsafe?"

78. Has any "intimidation" factor in communications been eliminated? Can anyone speak freely, to anyone else, about their honest safety concerns, without fear of career reprisals?

79. Do mechanisms exist that effectively promote and facilitate two-way communication between managers and all relevant stakeholders?

80. Is there a process to review the effectiveness of safety committees in promoting process safety and as a means to develop and execute a plan to improve such effectiveness?

81. Is the internal sharing of information that will reduce safety risks occur without fear of punishment?

82. Is there a strong emphasis on promptly recognizing and reporting nonstandard conditions to permit the timely detection of "weak signals" that might foretell safety issues? This issue is closely related to the normalization of deviance.

83. In general, do personnel not bother to report minor process-related incidents, accidents, or near misses?

84. Do several channels exist to communicate process safety critical information and to ensure that expertise can be accessed in a timely manner, especially in emergencies?

85. Does the organization interact with outside stakeholders regarding their hazards/risks?

86. Because of the legal concept of co-employment have host organizations consciously separated contractors, even

resident contractors, from certain host facility EHS activities so much that the resident contractors lack key safety or process safety information? For example, are resident and other contractors not allowed to attend host facility safety meetings, participate in host facility HIRAs/PHAs, or similar activities?

87. Is shift turnover a formal process? Is there a procedure or checklist for shift turnover? Is it logged?

Maintain a Sense of Vulnerability

88. Could a serious incident occur today, given the effectiveness of the current operating and process safety practices? When was the last serious close call or near miss?

89. Is there belief that compliance activities are guaranteed to prevent major incidents?

90. Are process safety policies, practices, and procedures institutionalized? Does the success of the PSMS rely primarily on the individual knowledge level, initiative, and decisions of those personnel who are assigned various responsibilities for process safety program elements and their activities?

91. Are lessons from related industry disasters routinely discussed at all levels in the organization? Has action been taken where similar deficiencies have been identified in the organization's operations?

92. Do hazard/risk analyses include an evaluation of credible major events? Are the frequencies of process safety events routinely determined to be unlikely and thus not credible?

93. Have proposed safety improvements been routinely rejected as not necessary because "nothing like this has ever happened here?"

94. Do risk analyses routinely eliminate proposed safeguards under the banner of "double jeopardy?"

95. Are critical alarms treated as operating indicators, or as near miss events when they are activated?

96. Is the successful functioning of a critical safeguard, e.g., a relief valve opens to relief overpressure, or a trip/interlocks functions at the correct setpoint to prevent a process safety incident regarded as a near miss or not?

97. Is the importance of preventive maintenance for safety critical equipment recognized, or is such work routinely allowed to be overdue?

98. Are the consequences of failure of critical equipment recognized and understood by all personnel?

99. Are there situations where the benefits of taking a risk are perceived to outweigh the potential negative consequences? Are there times when procedures are deviated from in the belief that major outcomes will not be caused? Are risk takers tacitly rewarded for "successful" risk taking?

100.   Is there an unreasonable "Can Do" attitude in place that minimizes risks and assumes that anything can be overcome (perhaps due to prior success recovering from a serious event)?

101.   Does the organization distinguish clearly between acceptable and unacceptable employee acts so that the vast majority of unsafe acts or conditions can be reported without fear of punishment?

102.   Is there institutional attentiveness to minor or what may appear as trivial/weak signals that may indicate potential problem areas within the organization and use of incidents and near misses as indicators of a system's "health?"

103.   Does the organization make weak responses to weak signals, i.e. do they set their threshold for intervening very high? If something does not seem right, they are very likely to continue operations and not investigate immediately? Does the organization have a low tolerance for "false alarms" and become de-sensitized to their occurrence?

104.   Do the explanations regarding the causes of incidents tend to be systemic rather than focusing on individual, "blame the

operator" justifications or other superficial causes (also referred to as reluctance to simplify)?

105.   Is the culture flexible so that it can adapt successfully to external influences, e.g., mergers/acquisitions, loss of key personnel, incidents, etc. without compromising safety?

106.   Is the organization resilient? That is, does the organization have the ability to recover from errors and return to normal operation despite challenges? This is characterized by a constant preoccupied with failure such that they can better anticipate areas of potential failure and can cope and bounce back from errors when they occur. This ability is in spite of their low incident rates.

107.   Does the organization fail to learn from past events?

108.   Has a pervasive attitude of complacency set in within the organization with respect to its hazards and risks? Does the organization lack a chronic sense of unease? For example, do they assume that because they have not had a process safety incident for ten years, one cannot happen imminently?

109.   Is process safety data gathering inadequate and does it focus on the wrong indicators or a limited set of indicators?

110.   Are there some individuals who have little interest in process safety? This can be a combination of the following factors: 1) the belief that process safety is somebody else's responsibility, 2) the concentration only on their own job and that process safety is something that other persons who are more knowledgeable will take care of, 3) lack of understanding of the process safety risks, and 4) lack of understanding of the integrated nature of the PSMS and its elements.

111.   Are process safety performance management, incentives and rewards related to a limited set of safety indicators (e.g. occupational injury rates) or not present at all?

112.   Is the control of risks reactive only?

113.   Do supervisors fail to perform frequent checks to confirm that workers (including contractors) are obeying safety rules?

114. Does the organization only seek information to confirm its superiority?

115. Does the organization believe that its process safety program has precluded process safety risk because it complies with regulations and standards?

116. Does the organization discount information that identifies a need to improve its process safety program?

117. Is there no interest in learning from other organizations or industries? Is the organization overly insular?

118. Are those who raise process safety concerns viewed negatively?

119. Does the response to process safety concerns focus on explaining away the concern rather than understanding it?

120. Are the investigations of process safety incidents superficial with a focus on the actions of individuals?

121. Are failures viewed as being caused by bad people rather than system inadequacy?

122. Is shift turnover a formal process? Is there a procedure or checklist for shift turnover? Is it logged?

123. Are operating procedures left to "gather dust on a shelf" because they are out of date, too cumbersome to use in day-by-day activities, or poorly written?

124. Have different MOC procedures/pathways been devised for different types of changes? This customizes the MOC requirements so that they are streamlined for particular purposes rather than relying one a single, comprehensive, and administratively burdensome "one-size-fits-all" MOC procedure.

125. Does the MOC process require proactive activities in advance of making a change, rather than the change occurring first and the MOC paperwork following merely as a form of documentation? This type of MOC process may leave behind a set of records that described what happened, but it obviates the entire purpose of MOC, which is to ensure that a proposed change receives a thorough and careful review and approval

before it is made. Establishing an MOC program only to comply with a regulation and leave a paper trail for each change does not fulfill the real purpose and intent of MOC. This is a form of complacency.

Understand and Act Upon Hazards/Risks

126. Does the organization know what standards govern the design, construction, maintenance, operations, and maintenance of its facilities? As with the boundaries, scope, and philosophy of application of the PSMS itself, has the organization taken a minimalist view of which recognized and generally accepted good engineering practices (RAGAGEP) applicable to the organization?

127. Does the organization follow its own procedures or does it regard them as not mandatory?

128. Are hazard/risk assessments performed consistently for engineering or operating changes that potentially introduce additional risks? Who decides if a risk assessment should be performed? What is the basis for not performing a risk assessment?

129. How are risks for low frequency – high consequence events judged? Is there a strong reliance on the observation that serious incidents have not occurred previously, so they are unlikely to occur in the future? What is the basis for deeming risks acceptable – particularly those associated with high consequence events?

130. Are the appropriate resources applied to the hazard/risk assessment process?

131. During HIRAs/PHAs are hazard scenario or type of hazard not included in study because it is bad news and will obligate management to do something tangible to reduce the risk, i.e., it will create a liability for management to spend resources to make the necessary changes to reduce the risk?

132. Do HIRA/PHA teams intentionally avoid making recommendations by applying risk rankings, IPL credits, or

other measurements of hazard/risk in a way that avoids the need for recommendations? This may be a cultural issue with a given HIRA/PHA team, but it may also represent a systemic problem in the organization at large.

133. Are HIRAs/PHAs performed "by the numbers" with little free thinking about what can go wrong? In recent years Layer of Protection Analysis (LOPA) has become a prevalent analytical method for determining "how safe is safe enough," particularly for high risk hazard scenarios identified during HIRAs/PHAs. LOPA has provided a consistent and repeatable method for determining how many independent protection layers are necessary to reduce the risk to a tolerable level. While LOPA is very useful analytical technique to analyze a hazard/risk, but it is not a very good technique for identifying a hazard/risk. Therefore, if the HIRA/PHA process at a facility relies solely on manipulating numerical credits to reach an acceptable cell on a risk matrix, without the cause and effect analysis and open discussion that occurs during HAZOP or What-If studies, the HIRA/PHA process may have lost an opportunity to identify additional important risks.

134. Are the recommendations emerging from the hazard/risk assessments meaningful? Do they address and reduce the risks identified?

135. Do risk reduction measures in HIRAs/PHAs over rely on human based safeguards such as operator training, the experience of personnel, or the existence of written operating procedures?

136. What are the bases for rejecting risk assessment recommendations? Are the reasons for rejection predominantly driven by cost considerations?

137. Are the risk assessment tools appropriate for the risks being assessed? Are the right tools to assess risks associated with low frequency – high consequence events? Are the tools deemed appropriate by recognized risk assessment professionals?

138. Is there a system, with effective accountabilities, for ensuring that recommendations from risk assessments are implemented in a timely fashion, and that the actions taken achieve the intent of the original recommendation?

139. Are the hazard/risk analysis performed as part of the MOC process adequate? Has this part of the MOC review process become somewhat pro-forma with little effort beyond routing the MOC to someone in the Safety Department for a routine review?

140. Are conflicts of interest allowed in the assignment of HIRA/PHA team leaders? For example, the process/project engineer who is responsible for the unit/system being studied should not lead the PHA on that process but should be a team member.

141. Does a questioning attitude prevail at all levels of the organization regarding the hazards/risks?

142. Are process safety risks and related controls communicated throughout the organization and beyond (contractors, other companies)?

143. Does management "face the facts" when necessary in response to process safety issues? Conversely, are difficult decisions regarding process safety issues routinely deferred hoping that the situation will be resolved in a different way?

144. Has the As Low As Reasonably Practicable (ALARP) principle been applied in making decisions about hazard/risk abatement? Has the ALARP principle been applied reasonably and consistently?

145. Have formal definitions of tolerable risks that have been agreed-to by the entire organization been adhered to without regard to their ramifications? For example, if a risk based inspection (RBI) program has been implemented have the ITPM frequencies that allow the process safety risk to remain at a tolerable level been followed even if this requires that equipment be shutdown unexpectedly to perform a needed test or inspection?

146. Have the results of hazard/risk analyses been used to plan, organize, and execute the other elements of the PSMS? Examples: use of the causes and safeguard information developed during HIRAs/PHAs to determine which equipment should be included in the AI/MI program, the use of HIRA/PHA results to determine process safety related training for operators, maintenance, and other personnel, the use of HIRA/PHA results to determine the contents of the emergency response plan?

147. Has the organization included analysis of inherently safer design (ISD) considerations in its process safety program? Are appropriate ISD provisions implemented when feasible?

148. (U.S.A.-specific) Has the organization interpreted what constitutes offensive vs. defensive actions when trying to determine whether a U.S. facility is responding to emergency events at a level that would invoke the HAZWOPER (1910.120) regulations? This is affected to a large degree by the emergency response culture that has been established within the emergency response team at the facility, and by the philosophy that was used to develop and emergency provisions of the operating procedures.

149. Have key leading and lagging process safety metrics been established and reported to management on a periodic basis? Are the process safety metrics defined in such a way as to artificially indicate a PSMS status that is not completely accurate. For example, overdue ITPM metrics that include only those from the main MI/maintenance data base and do not include all of the process safety-relevant ITPM tasks being performed (e.g., fire protection equipment managed separately)?

150. Do process safety metrics never vary from very high/positive values? While this may seem satisfying, it usually does not comport with the reality of actual facility operations.

151. Are PSMS audits and process safety metrics met by severe pushback?

152. Do personnel seem to focus much more on the scores of audits (if they are scored in any way) and on the value of process safety metrics and not on the facts and findings that underlie them?

Empower Individuals to Successfully Fulfill their Safety Responsibilities

153. Are there means to develop and implement new plant-level process safety goals, policies, practices, and procedures that take into account the interests and input of relevant internal stakeholders?

154. Is the company an industry leader in process safety by taking a leading role in industry process safety organizations and activities and sharing results and learnings with the industry? Examples of such organizations include CCPS, MKOPSC, DIERS, API, NFPA, etc.

155. Is the model for controlling process safety within the organization centralized or decentralized? Is this model appropriate given the personnel, resources, and documentation methods available? Is the model currently in place sustainable?

156. Do the facility personnel in key process safety roles have the prerequisite knowledge and skills? This goes beyond simple awareness. For example, if the organization facilitates its own HIRAs/PHAs, then the internal facilitators should be subject to a formal internally defined training and qualification program, and the facilitators should have successfully completed that program before they are assigned the lead HIRAs/PHAs by themselves. If the facility is subject to corrosion or damage mechanisms that are complex or not typical, then either facility, company, or outside corrosion engineering or metallurgy expertise should be available for consultation, inspections, or other technical work on an as-needed basis without undue delay.

157. Is organization senior management competent in the technical aspects of process safety and how process safety regulations are interpreted for the site/company beyond a simple awareness level?

158. Is middle management, including EHS managers and the process safety Manager/Coordinator competent in the technical and regulatory aspects of process safety?

159. Do personnel with support or peripheral roles in the PSMS understand the technical aspects of process safety, commensurate with and as it applies to their jobs?

160. Is the hourly workforce competent in the technical aspects of process safety, as it applies to their jobs?

161. Is there is a continuing process safety training and education curriculum for all personnel, appropriate to their responsibilities and roles in the PSMS?

162. Are employees are empowered to suggest or initiate improvements?

163. Does the process safety training allow personnel to recognize when a process should be shut down if safety critical interlocks, alarms or other process-safety devices fail or become unavailable during operation?

164. Are operators empowered to take corrective action as soon as possible (including shutting down when appropriate) if safety critical interlocks, alarms, or other process safety-related devices fail or become unavailable during operation?

165. Is there built-in redundancy in the processes, their procedures, and practices? Examples include: providing back-up systems in case of a failure, internal cross-checks of safety-critical decisions and continuous monitoring of safety critical activities, e.g., the use of the "buddy system" whereby activities carried out by one individual are observed by a second member of staff.

166. Is decision-making hierarchical during routine periods, accompanied by a clear differentiation of responsibilities? In emergencies, does decision-making migrate to individuals

with expertise irrespective of their hierarchical position within the organization?

167. Do managers monitor decisions but do not intervene unless required, usually when there is an unplanned deviation in a course of action?

168. Are there well-defined procedures for all of the possible unexpected events that have been identified?

169. Does a climate of continuous training exist in order to enhance and maintain operator's knowledge of the complex operations within the organization, improve their technical competence and enable them to recognize hazards and respond to 'unexpected' problems appropriately?

170. Does the organization over-rely on computer based training (CBT) for EHS training? Is the quality of training delivered via CBT appropriate?

171. Do the relevant employees participate in the full spectrum of process safety management activities, including the setting safety standards and rules? Do they participate in both the development and implementation of the PSMS and activities where they have interest and where they have concerns?

172. Have organizational "silos" been broken down with respect to the PSMS? "Silos" refer to organizational barriers that inhibit the free exchange of information and ideas and can also inhibit collaboration and cooperation.

173. Is a person's process safety performance is considered when hiring, retention, and promotion decisions are being made?

174. Do positive labor relations exist (if a union represents some or all of the workers)?

175. Are those with responsibility for representing employees (e.g. health and safety committee members) provided with adequate training, skills, and resources?

Defer to Expertise

176. Does the organization defer to expertise in identifying, evaluating, and controlling the hazards/risks? Are there

competent PHA/HIRA team leaders available to consistently, completely, and accurately perform these studies?

177. Are the leaders and other persons of influence in the PSMS with the appropriate expertise skilled in the hazard recognition for process safety? Are these person's opinions given the appropriate value in the debate of process safety issues and decisions?

178. Is the expertise of qualified operators deferred to regarding emergency shutdown decisions without second guessing?

179. Do people who perform or conduct process safety activities have the appropriate competence to interpret and apply the underlying process safety requirements?

180. Are the persons who lead or conduct investigations of process safety incidents or near misses skilled in the root cause analysis techniques employed?

181. Are the persons who lead or serve as PSMS auditors skilled in this activity, including internal auditors?

182. Are personnel with the required and correct expertise placed in the untenable position of proving that their recommendations and suggestions for decision or action on process safety issues are in the unsafe direction rather than those who hold opposing views are in the safe direction?

183. Are personnel with the required and correct expertise able to influence events when required? For example, can they delay the approval of a desired (and rushed) MOC? Can they delay a startup that is anxiously anticipated during a PSSR? Are their interpretations valued heavily in debates regarding PSMS decisions?

184. Has the appropriate expertise for complex AI/MI corrosion and damage mechanisms been obtained? Is this expertise applied when decisions regarding the inspection and testing of fixed equipment, and the removal from service when necessary?

185. Are the decisions during MOC reviews being made by the right people, i.e., do these reviewers/approvers of MOC

understand the technical issues involved? For example, is the person(s) making decisions regarding what type of safety review (a simple checklist review vs. a PHA) is necessary for each MOC competent to make this decision?

186. Are conflicts of interest for various process safety activities avoided if possible? For example, is the originator of an MOC allowed to give the final approval for it? Is the originator of an MOC allowed to decide what level of safety review should be performed for that change?

187. Are the personnel with appropriate expertise in control of or have significant influence over the PSMS activities where their expertise is necessary, i.e., are those people empowered to influence the activities and decisions in the area(s) where their expertise applies?

188. Do managers have the appropriate technical competence/expertise to make key process safety decisions? If not, is this expertise available to them directly?

189. Is the proper expertise for key process safety related decision making applied, e.g., approval of safe work permits, approval of bypass of safety features, etc.?

190. Do organization personnel understand and can they successfully apply the codes, standards, and other written guidelines that constitute the RAGAGEPs that govern the design, construction, ITPM, and operation of the organization's operations? If the organization has this expertise, are the people who have it in control of or have an influence over the PSMS activities where the RAGAGEPs apply?

Combat the Normalization of Deviance

191. Has the organization taken such a minimalist view of which recognized and generally accepted good engineering practices (RAGAGEP) applicable to the organization that there are gaps between what the RAGAGEPs require and what the organization has included from them?

192. Are there systems in operation where the documented engineering or operating design bases are knowingly exceeded, either episodically, or on a "routine" basis?"

193. Have there been operating situations where problems were solved by not following established procedures, or by exceeding design conditions? Does the organizational culture encourage or discourage "creative" solutions to operating problems that involve circumventing procedures?

194. Do organization personnel believe that MOC is important, and that changes cannot occur, however convenient they may be, or however simple and obvious they may seem without the appropriate review and authorization?

195. Is it clear who is responsible for authorizing waivers from established procedures, policies, or design standards? Are the lines of authority for deviating from procedures clearly defined? Is there a formalized procedure for authorizing such deviations?

196. What action is taken, and at what level, when a willful, conscious, violation of an established procedure occurs? Is there a system to monitor deviations from procedures where safety is concerned? Can staff be counted on to strictly follow procedures when supervision is not around to monitor compliance?

197. Are the management systems sufficiently discerning and robust to detect patterns of abnormal conditions or practices before they can become accepted as the norm?

198. Are systematic analyses of incidents conducted to identify their root causes and accident types or trends within the organization?

199. Does the facility clearly distinguish between acceptable and unacceptable employee acts so that the vast majority of unsafe acts or conditions can be reported without fear of punishment? For example, is there an attitude that small fires are considered commonplace and are a "fact of life" in the plant and therefore can be tolerated as is?

200. Are check-the-box activities tolerated? Check-the-box activities those that are performed to simply complete them, regardless whether the activity was performed and documented properly and thoroughly and anything was actually learned from it (i.e., simply checking it off the list of things to do). Examples: a full PSMS audit of a large facility such as an oil refinery that took only 1 day to complete with a very small audit team, or a HIRA/PHA of a major project that was completed in only one session.

201. Are process safety problems resolved and corrective actions implemented in a timely manner? In this context, "timely" means that resolution or corrective action plans are promptly determined, the recommendations are resolved quickly, and the implementation of the final action is completed in a time period that is reasonable given the complexity of the action and the difficulty of implementation. The timing of resolution plan development and completion of each recommendation should be evaluated on a case-by-case basis.

202. Is there a formal deferral process for PSMS action items or activities that are time-based? If so, how many deferred tasks or activities are there and what is their aging? Has the formal deferral process masked a large number of overdue tasks/activities by simply re-classifying them from "overdue" to "deferred?"

203. Is a priority placed on the timely communication and response to learnings from incident investigations, audits, HIRAs, etc.?

204. Are discrepancies between practices and procedures (or standards) resolved in a timely manner to prevent the normalization of deviance?

205. Do supervisors sometimes asks operators to operate equipment that is in an unsafe condition? Will supervisors support operators if they refuse to participate in unsafe work?

206. Do supervisors take action when a worker engages in a poor process safety practice?
207. Are disabled or failed process safety devices restored to service as soon as possible?
208. Are interlocks, alarms, and other process safety-related devices are regularly tested and maintained?
209. Are written operating procedures are regularly followed?
210. Do personnel sometimes work around process safety concerns rather than report them? Are unapproved shortcuts around process safety tolerated?
211. Have the following indicators of the normalization of deviance appeared in the organization and its operations:
    a. Safety systems/features that remain removed from service beyond the time limits specified, or removals that are continually extended and become quasi-permanent, even if these extensions are in accordance with policy or procedure.
    b. Alarms that are constantly sounding (i.e., nuisance alarms). Rather than pursuing an alarm management program to improve the alarm system, the alarms are simply ignored because personnel come to believe that they represent nuisance conditions rather than actual process deviations.
    c. Operators do not believe their indications because the instrumentation is chronically not calibrated or inaccurate and hence personnel are hesitant or resistant to taking firm actions based upon them such as emergency shutdown.
    d. Chronically overdue ITPM tasks in the MI/AI program, or a list of the overdue ITPM tasks that is growing and the aging of the overdue tasks is lengthening.
    e. Is there a formal deferral process for ITPM tasks at the facility? If so, how many deferred tasks are they and what is their aging? Has the formal deferral process masked a

    large number of overdue ITPM tasks by simply re-classifying them from "overdue" to "deferred?"

f. Growing lists of equipment deficiencies and increasing aging of these deficiencies (i.e., they have been open for lengthy periods of time).

g. Not extending a turnaround for a few days when additional work is needed to perform work that is critical to process. It may be necessary to delay the startup of a unit to accomplish crucial tasks that the facility may not be able to do until the next turnaround, which might be several years in the future. This will make certain ITPM tasks or the correction of important AI deficiencies overdue. Examples: 1) Not having time to remove all temporary clamps from leaks in piping by installing the preplacement piping circuits. 2) Not performing all required SIS proof tests or other instrumentation or safety device tests that require the unit or equipment to be shutdown. It will take a courageous facility manager to recommend to senior organization management that a planned turnaround be extended, even by a few days. However, sometimes it is absolutely necessary.

h. The adoption of programs that establish a tolerable risk and set the frequency for ITPM based to stay within the tolerable risk boundaries, and then ignoring the ITPM frequencies necessary to remain within a tolerable risk. Examples: RBI (voluntary) and the SIS Standard (required by RAGAGEP). When scheduled turnarounds are arbitrarily extended, the RBI or SIL calculations used to establish the appropriate ITPM frequencies are invalidated.

i. The resolution of PSMS related action items from HIRAs/PHAs, incident investigations, audits, and other activities is chronically overdue and the aging of open recommendations and action items keeps lengthening.

j. Operators do not follow approved operating, maintenance, or other procedures, particularly when supervision is not present. Also, there is toleration of short-cuts in operations or maintenance tasks. There is the use of peer pressure to help force members of the workforce or newcomers to take the short-cuts or not follow required procedures verbatim is an especially insidious practice.

k. Allowing conflicts of interest to persist in the management of PSMS elements. For example, allowing the initiator of a proposed change to approve the MOC, or allowing the initiator of a proposed change to perform the analysis of the impact of the change on safety or process safety during the MOC review process. Other examples include allowing Operations to make the final decisions regarding the formal approval of ITPM deferrals, or allowing the process engineer with the main responsibility for engineering, design, and project issues in a given process to lead the HIRA/PHA for that process.

l. Operator rounds become pro-forma activities and merely an exercise to record certain data rather than a careful, thorough, and critical examination of the operating status of the equipment. Training operators to be sensitive to changes in noise, smell, and other sensatory inputs will help discover problems before they become serious. Ignoring these changes in equipment characteristics normalizes them.

m. Performing shift turnover in an incomplete or casual manner, e.g., turnover between operators conducted in the locker room rather than in the unit or control room, or no turnover at all.

n. Allowing relatively minor issues, e.g., poor housekeeping or steam/water leaks to persist. While the lack of cleanliness may only be an eyesore in some cases, it fosters an attitude that management does not care and over time it erodes the sense of pride in the facility. Eventually poor housekeeping will be a cause or a contributor to a safety or process safety incident. Housekeeping, in a broad sense, also includes the condition of paint, insulation, lighting, as well as complete and accurate equipment labeling, although special emphasis programs may be established for these issues.

o. Artificially risk ranking HIRA/PHA or SIL assessment scenarios lower than they should be to avoid the requirements for more layers of protection/safeguards.

p. Stretching the definition of "annual," e.g., doing something required on January 1st of one year and again on December 31st of the following year and still considering having met an annual requirement.

q. Not managing all changes in a formal manner (i.e., using MOC) because it is considered an administrative burden. For example, allowing small changes that warrant MOC to be made without MOC such as classifying valve type changes (e.g., globe to gate, etc.) as a "P&ID change" or replacement-in-kind.

r. Not consistently investigating near misses thoroughly and formally. The formal and thorough investigation of near misses is an excellent opportunity to learn something without suffering any of the bad consequences. Also, the narrow or minimalist definition of near miss is a normalization of deviance, e.g., not investigating the root causes of why a safeguard operated as designed. If a trip or interlock operates successfully at its setpoint is classified not as a near miss but as a designed or intended operation, the root cause or underlying condition that caused the safeguard to function can then become normalized. It misses the opportunity (and need) to know why the safeguard was challenged, which was not intended.

s. Not investigating a minor incident or near miss because it is bad news and will obligate management to correct it.

t. Safe Work Practices that evolve into purely administrative exercises, or nearly so. Allowing safe work permits to be consistently completed in a haphazard manner where required fields are chronically left blank, signatures are missing, and the permits are not closed out properly. Also, allowing SWP programs to be paper-only programs with no field verification of actual conditions by those responsible for inspecting the work locations before approving the permits.

u. Oversimplification of process safety risks. In this context, oversimplification means evidence of some risks is disregarded or deemphasized while attention is given to a handful of others. The reluctance to simplify risks results from systematically collecting and analyzing all warning signals and avoiding making assumptions regarding the causes of failures; and using accident investigations to identify the potential systemic factors contributing to incidents.

v. Required process safety related training activities that are chronically overdue.

w. Operating procedures that are not controlled documents and are not up to date.

x. The use of a separate training manual rather than the SOPs themselves in the operator training program. This indicates that the SOPs are not very well written. Also, the operators will tend to rely on a well-written training manual as an operating reference rather than the officially approved SOPs.

y. MOC reviews that are overdue and backlogged resulting in lengthening lists of MOCs that have not been completed (as MOCs age the validity of the reviews that were performed to approve them erodes).

z. Allowing temporary changes to become quasi-permanent due lack of periodic management review and challenge. The MOC procedure of an organization should specify a hard cap on how long a temporary MOC may exist. If additional time is needed and it is warranted, then the procedure should contain provisions for waiving the cap following appropriate review and approval.

aa. Using emergency/verbal MOCs sparingly? Most MOC procedures allow for the rapid review and approval of a change when the circumstances warrant, e.g., a change that is needed quickly for safety or other reasons during off-hours. The steps covering this type of rapid approval vary, but they usually involve obtaining a verbal approval from one or more designated persons, often over a telephone. The normal MOC forms and process are then completed at the next normal work time availability. Clearly, this is a simpler and quicker way of getting an MOC approved than the normal process and because of its convenience can be abused. At its core, an MOC procedure is a sophisticated and formal communications system where designated persons render opinions (i.e., review) or make decisions (i.e., approve) about a proposed change. It simultaneously serves as the documentation method for this communications process.

bb. Applying to selected personnel and organizational changes. Not all personnel and organizational changes should warrant the use of MOOC and it is currently the choice of each organization what types of personnel changes will be included. At a minimum, those positions that are critical to the PSMS within an organization, or personnel actions that can have a significant effect on the PSMS should be included within the scope of OMOC. Of course, the details of job duties and responsibilities in a particular organization are crucial in making such decisions, and all organizations are not identical with respect to assigning duties and responsibilities.

cc. The management of "red lined" documents is chronically behind. "Red line" documents are temporary marked-up copies of P&IDs and other drawings, operating procedures, or other documents that are created pursuant to making a change and are maintained for Operations, Maintenance, and other personnel while the master versions are being permanently revised. The red-lined documents are usually made available in the control room, on a network if they have been scanned, or other convenient location(s). There is a tipping point when the amount of out-of-date information becomes overwhelming and a major upgrade program, with substantial, set aside resources in of time, money, and personnel is necessary to bring the information back to an accurate, useful state.

dd. The Operational Readiness review/PSSR process is compromised in the rush to startup.

ee. Relief system/device design that is not analyzed and up-to-dated as products or operating conditions change. Assuming that the previous installed device will "envelope" the conditions imposed by a new product is not the correct technical approach. Neither is asking a few simple questions during a qualitative hazard analysis for the product change and presuming that this level of analysis is sufficient. Confirmation that the reaction kinetics and thermal/pressure characteristics of a new product are fully protected against is an important process safety activity.

ff. "Gamesmanship" in the collection of process safety metrics occurs. Since process safety metrics are usually defined in numerical terms, there will be an incentive to make the numbers look as positive as possible. This is one of the human influences on this activity, and is a natural human inclination.

gg. Not including a finding/observation in a PSMS audit because it is bad news and will obligate management to do something tangible to correct is an indicator of a negative process safety culture.

212. Does the organization fail to implement or consistently apply its management system across the operation (regional or functional disparities exist)?

213. Are safety rules and defenses are routinely circumvented in order to get the job done, or done more quickly and cheaply?

214. Does the organization failing to provide necessary financial, human, and technical resources?

215. Are there impractical process safety rules, processes and procedures, which make compliance and achievement of other organizational outcomes mutually exclusive?

216. Do personnel find workarounds in response to operational inadequacies?

217. Are operational deviations managed without using change and risk management processes?

218. Is there is an extended time between reporting of safety issues (hazards, inspection and audit findings, other deficiencies, etc.) and their resolution?

219. Has the organization re-written their safe work practice (SWP) procedures to reinforce them (or a subset of them) as inviolable "lifesaving rules?" If so, do they enforce them as such?

220. Is there undue peer pressure or organizational pressure to work extensive amounts of overtime?

221. Does the culture of different operating shifts within a given facility vary widely? Most of this difference can be attributed to the attitudes and beliefs of supervisors, foremen, and other mid-level management assigned to off shifts. Having to accomplish the same level of production with less resources sometimes fosters the attitude that shortcuts and other

deviations from approved procedures are acceptable in order to achieve those goals. Sometimes this situation fosters a "can-do" attitude where independently finding any solution to a problem and getting around a production delay/issue quickly is seen as a positive attribute.

222. Is fitness for duty an issue within the organization? No external influences on performance should be allowed for facility personnel, particularly operators. This includes the use of alcohol or drugs while on duty, horseplay, harassment, and other aberrant behaviors. No tolerance for these behaviors goes beyond simply forbidding them in written policies and stating these prohibitions during training/orientation sessions. It means living them at each level of the organization, including self-policing within peer groups and during off shifts. Fitness for duty also includes issues such as fatigue and shift rotation schedules.

Learn to Assess and Advance the Culture

223. Do leaders consistently model and support the attitudes and behaviors that are expected of the culture? Do the workers emulate them?

224. Has the organization established and nurtured a questioning and learning environment?

225. Is the organization hesitant to share information and lessons learned both inside and outside the organization?

226. Are there systems in place for reliably learning from our mistakes? Does the organization willingly and enthusiastically accept those learnings and apply them to improve process safety systems and procedures?

227. Is there a plan to improve the process safety culture?

228. Are anonymous process safety culture surveys conducted periodically to measure the effectiveness of efforts to improve process safety culture?

229. Have key process safety culture metrics been established and reported to management on a periodic basis? Review recent

process safety culture metrics and their reporting and follow-up.

230. Does the company board of directors monitor the status and progress of the company's PSMS? If the company is not publicly traded and no board of directors exists, does the owner(s) or those designated by the owner(s) perform this role?

231. Is the organization adaptable from a process safety viewpoint? Is the working environment is characterized by acceptance to change? Indicators of resistance to change include: the "not invented here syndrome" where changes to procedure and policy are not accepted because they were developed somewhere else or by someone else; bureaucratic inertia where too many people have to implement a programmatic change; inadequate training and explanation of the changes (especially why the change is necessary).

232. Is the organization adaptable? Adaptability means that the organization's overall culture helps them manage internal change well. These organizations are very responsive and incorporate change easily, such as a change of CEO. Adaptable organizations continually adopt new and improved ways to do work, and the different units or groups in these organizations often cooperate to create change.

233. Does the organizational culture stress team orientation? Teams are the primary building blocks of team-oriented organizations. Cooperation and collaboration across cross-functional roles are actively encouraged in organizations. Work is sensibly organized in these organizations so that each person can see the relationship between his/her work and the goals of the organization.

234. Have the facility personnel who do not hold key process safety roles been thoroughly indoctrinated into the precepts of process safety and how it applies at the facility? Are there positive signs that the greater facility population understands and appreciates what process safety is, why it is necessary,

and how they fit into and support the program? In other words, simply requiring all personnel to periodically sit through a CBT module on process safety is usually not enough. An increase in the number of near misses reported, improved housekeeping, decreasing reluctance to challenge status-quo practices and habits are all signs that the facility population at large has started to absorb both the facts associated with the process safety program, as well as its intended spirit.

235. Does a learning environment exist where not just lots of training is mandatory, but also where there is widespread use of the word "why" by facility personnel? Natural curiosity is a human character trait and some people have more of it than others and are willing to indulge that curiosity as part of their jobs. Others are more reticent. A positive process safety culture will encourage and satisfy the natural level of curiosity of the facility staff and may also improve it.

236. Does the organization have both personal and organizational antennae that are sensitive enough to detect telltale signs of impending process safety problems before they become incidents? This means being able to sift through a large amount of formal and informal information for the "nugget" of intelligence that portends a problem. In the post mortem of nearly all undesired events, the investigations reveal that the information needed to detect, prevent, or mitigate the events in question were available to the organization but their context was not understood or the information was ignored. Organizations that develop "20/20 foresight" possess an invaluable tool for preventing incidents. This trait should be firmly encouraged and rewarded when it is displayed. One aspect of developing such a trait is to always thoroughly investigate process safety near misses. This should be done not because it is a regulatory requirement in many places but because it represents a golden opportunity to learn lessons without suffering severe consequences. This

is a difficult thing to accomplish for several technical and cultural reasons: it is hard to understand the nuances of many process safety incidents and outside assistance may be necessary to do a thorough job, and few people enjoy the process of a formal investigation where despite best attempts to avoid it, some blame is often apportioned. However, when people are truly curious about why something has happened and investigating their root causes is not regarded as a burden but as opportunities to learn, that desired "20/20 foresight" will start to develop. Another aspect of developing such a trait is to collect a large amount of information from multiple sources and process it thoroughly via a technically valid and tested vetting process. Also, in the collection of this information, the messenger is never punished and conversely, personnel are encouraged to raise issues without fear of either retribution or being marginalized as always "crying wolf."

F.3 References

F.1   Baker, J. et al., The Report of BP U.S. Refineries Independent Safety Review Panel, January 2007 (Baker Panel Report).

F.2   Contra Costa County (CCC) Industrial Safety Ordinance, County Ordinance Chapter 450-8 (as amended).

F.3   Center for Chemical Process Safety (CCPS), Guidelines for Auditing Process Safety Management Systems, 2nd Ed., American Institute of Chemical Engineers, 2010.

F.4   Canadian National Energy Board (CNEB), Advancing Safety in the Oil and Gas Industry - Statement on Safety Culture, 2012.

F.5   Jones, D., Frank, W., Tancredi, K., Broadribb, M., Key Lessons from The Columbia Shuttle Disaster (With Adaptation to The Process Industries), AIChE, 2006.

F.6   UK HSE, Human Engineering, A review of safety culture and safety climate literature, Research Report 367, 2005.

F.7   UK HSE, Development and validation of the HMRI safety culture inspection toolkit, Research Report 365, 2005.

APPENDIX G: PROCESS SAFETY CULTURE & HUMAN BEHAVIOR

The full reference may be downloaded from www.aiche.org/ccps /publications/guidelines-culture.

Essential Practices for Creating, Strengthening, and Sustaining Process Safety Culture, First Edition. CCPS. © 2018 AIChE. Published 2018 by John Wiley & Sons, Inc.